AaBbCcDdEeFfGgHhIiJjKkLlMmNn

Twenty Years Before the Blackboard

©1998 by
The Mathematical Association of America (Incorporated)
Library of Congress Catalog Card Number 97-74343
ISBN 0-88385-525-9
Printed in the United States of America
Current Printing (last digit):
10 9 8 7 6 5 4 3 2 1

Twenty Years Before the Blackboard

The Lessons and Humor of a Mathematics Teacher

Michael Stueben
with
Diane Sandford

Published by
THE MATHEMATICAL ASSOCIATION OF AMERICA

SPECTRUM SERIES

The Spectrum Series of the Mathematical Association of America was so named to reflect its purpose: to publish a broad range of books including biographies, accessible expositions of old or new mathematical ideas, reprints and revisions of excellent out-of-print books, popular works, and other monographs of high interest that will appeal to a broad range of readers, including students and teachers of mathematics, mathematical amateurs, and researchers.

All the Math That's Fit to Print, by Keith Devlin
Circles: A Mathematical View, by Dan Pedoe
Complex Numbers and Geometry, by Liang-shin Hahn
Cryptology, by Albrecht Beutelspacher
Five Hundred Mathematical Challenges, Edward J. Barbeau, Murray S. Klamkin, and William O. J. Moser
From Zero to Infinity, by Constance Reid
I Want to be a Mathematician, by Paul R. Halmos
Journey into Geometries, by Marta Sved
JULIA: a life in mathematics, by Constance Reid
The Last Problem, by E. T. Bell (revised and updated by Underwood Dudley)
The Lighter Side of Mathematics: Proceedings of the Eugène Strens Memorial Conference on Recreational Mathematics & its History, edited by Richard K. Guy and Robert E. Woodrow
Lure of the Integers, by Joe Roberts
Magic Tricks, Card Shuffling, and Dynamic Computer Memories: The Mathematics of the Perfect Shuffle, by S. Brent Morris
Mathematical Carnival, by Martin Gardner
Mathematical Circus, by Martin Gardner
Mathematical Cranks, by Underwood Dudley
Mathematical Magic Show, by Martin Gardner
Mathematics: Queen and Servant of Science, by E. T. Bell
Memorabilia Mathematica, by Robert Edouard Moritz
New Mathematical Diversions, by Martin Gardner
Numerical Methods that Work, by Forman Acton
Numerology or What Pythagoras Wrought, by Underwood Dudley
Out of the Mouths of Mathematicians, by Rosemary Schmalz
Penrose Tiles to Trapdoor Ciphers . . . and the Return of Dr. Matrix, by Martin Gardner
Polyominoes, by George Martin
The Search for E. T. Bell, also known as John Taine, by Constance Reid
Shaping Space, edited by Marjorie Senechal and George Fleck
Student Research Projects in Calculus, by Marcus Cohen, Edward D. Gaughan, Arthur Knoebel, Douglas S. Kurtz, and David Pengelley
The Trisectors, by Underwood Dudley
Twenty Years Before the Blackboard, by Michael Stueben with Diane Sandford
The Words of Mathematics, by Steven Schwartzman

MAA Service Center
P. O. Box 91112
Washington, DC 20090-1112
800-331-1MAA   FAX 301-206-9789

Preface

This book is the legacy of my professional life as a high school mathematics and computer science teacher. It contains truths about teaching high school mathematics that took me nearly two decades to discover. It is also a scrapbook of the best mathematical humor, wordplay, and curiosities that I have encountered during my career. As the mathematician J.E. Littlewood once said, "A good mathematical joke is better, and better mathematics, than a dozen mediocre papers."

—Michael Stueben (December 1997)

This book is affectionately dedicated to my colleagues and former students. I learned more from you than you learned from me.

Acknowledgments

I would like to thank the following people who have aided me in research and encouragement: Martin Gardner, Donald J. Albers, Underwood Dudley, David Singmaster, Howard Eves, Peter Renz, Daniel A. Asimov, Isaac Asimov for his biography, Scot Morris of *Omni*, James Stanlaw of Illinois State University for his lucid explanations of society, Malcolm K. Smithers of London for his enlightening correspondence, Lee Sallows of The Netherlands for his charm and ingenuity; my teaching colleagues Sally Bellacqua, Jerry Berry, André Samson, Dennis McFaden, Steve Rose, and Fran Salerno (all master teachers who have shared their teaching philosophies with me); and my good friend in mathematical recreations Michael H. Brill. But most of all, I would like to thank my wife, Diane Sandford, who supported me for a year of research, and who has spent countless hours trying to teach me the craft of writing.

I wish to thank the following publishers for their generous permission to reprint certain quotations included in this text.

Chapter 1, page 8: From *Out of My Later Years* by Albert Einstein. Copyright ©1956, 1984 by the Estate of Albert Einstein. Published by arrangement with Carol Publishing Group.

Chapter 1, page 13: Roy Dubisch, *The Teaching of Mathematics* (John Wiley, 1963), Copyright ©(John Wiley & Sons, 1963). Reprinted by permission of John Wiley & Sons, Inc.

Chapter 1, page 18: Reprinted from "Once Over Lightly," by J.L. Kelley, *A Century of Mathematics in America, Part III,* Peter Duren, Ed, pp. 474–475, 1989. Reprinted by permission of the American Mathematical Society.

Chapter 2, page 34: *Etc.: A Review of General Semantics*/vol. XVI, no 4, Summer 1959/pages 457–458.

Contents

AaBbCcDdEeFfGgHhIiJjKkLlMmNn

Part 1

Teaching

1
Twenty Years Before the Blackboard

He came from the Town of Stupidity,
which lieth four degrees to the Northward
of the City of Destruction.—John Bunyan,
The Pilgrim's Progress Part II (1684).

Lesson 1: The Students

In 1931 an obscure secondary school teacher named William H. Patterson
wrote a book called *Letters From A Hard-Boiled Teacher To His Half-Baked
Son.* This book contains letters written by Patterson, a teacher with forty years
experience, to his son upon the son's graduation from a teachers college.
It begins: "If ignorance of the real tricks of the teaching profession were a
crime, you should be shot at sunrise." Here, in two paragraphs, the old teacher
explains a major truth in teaching:

> You will find no better motto at this stage than "Learn Your Pupils."
> Both your Alma Mater and the system in which you teach emphasize
> the head at the expense of the heart. Learning is achieved, but the
> pupils themselves, their likes and dislikes, their preferences, tastes,
> secret desires, opinions, and affections are neglected. Influenced by
> your environment you are in danger of thinking of your pupils as
> simply a part of the machine or a certain amount of material. I want
> to warn you against overlooking the personal relationship between
> your pupils and you which is one of the most important factors
> in teaching. Beware of drifting into that large group of teachers
> who are more interested in the stuff to be taught than in children,

who have no time to waste in getting personally acquainted with pupils. A teacher with such an attitude can never make a fine art of teaching.

Perhaps the books, teachers, and schools have not so filled you with their picture of a rational humanity that you will be 40 years in learning the truth; but that is what they did to me. If your pupils like you, no matter how weak and shallow you are, they will support you; and if they don't like you, no matter how wise and good you are, they will not follow you. But before you can induce your pupils to like you, you must understand them, and you can't understand them until you study and learn them.

Nowhere have I seen this idea more forcefully stated than in Patterson's book. When I began to teach in 1975, I did not have Patterson's perspective and would have doubted it if someone had explained it to me. It took me fifteen years to discover his truth.

Lesson 2: Compliments

One of the greatest revelations I ever had in teaching was the realization that compliments were a serious business. Years ago I saw a movie about a newly married wife who bought a book called *How To Train Your Dog* and applied the principles to her husband. One of the principles was to reward the dog every time he answered to his name. So every time the young wife needed her husband, she called him and then rewarded him with a kiss, food, or a compliment. It worked perfectly until he found the book and realized his wife was using a form of mind control. That may have been fiction, but rewarding someone for good behavior is an effective idea.

At the beginning of the year, most of my classes pay attention to me. I need them to do this all year, so I mention their behavior and tell them how nice it is that they listened, and that this was a fun class to teach today, and that their listening skills were better than many classes I have had in the past, etc. I compliment the actions I want them to repeat.

My father had a great trick. Nearly each day he would ask my mother if he had told her recently that he loved her. She would say no, and then he would launch into a short lecture about her good points, or his inability to appreciate her, or something nice about her that was fun to hear. After a while it became a game they played. But it was a game that required him to compliment her nearly once a day. Why did he play it? Because he knew compliments are necessary to maintain a good relationship.

Lesson 3: Whining

Whining is complaining that the world is not as we wish it to be. I know two teachers who have signs on their desks saying "No Whining." I thought this was pretty funny the first time I saw it. Student whining rarely annoys me. What really annoys me is my own whining.

A student once asked me to look over his computer program to help him find an error. I sat down and took a look. He did not indent well, the subroutines were too big, and the identifiers were non-descriptive. With those kinds of errors, I had no choice but to start at the beginning and verify each block of code. I immediately discovered his first procedure had a serious error: It did not load in the proper data from a file. He had violated the most important rule of programming: Do not write a new procedure until you verify the current one. This was too much frustration, and I hit the roof.

> "You did this wrong. You did that wrong. Of course your program doesn't work, because you're not writing it in such-and-such a way. The reason I told you to adopt this style is so that you won't have these problems. Blah, blah, blah."

I finally ended by telling him that instead of learning "programming" from me, he had learned "non-programming" and had developed such counter-productive skills that he was worse off than somebody who had never taken the course. Did I really say that to a student? Unfortunately, I did. The issue here is not whether I was a poor teacher or whether he was a poor student. The issue is how my ego handled the shock of discovering what he had learned under my influence. I criticized, complained, blamed, and whined. The result was that nothing changed except his opinion of me and maybe his self-respect. The way out of this was mess was to apologize to the student and to try to follow the big rule: *Thou shalt not whine.*

Mistakes in the classroom have been ideal teaching moments for me. These mistakes tell me my methods are wrong or that I am not paying enough attention to my students. The problem is that mistakes are painful to admit. It's easier to blame the student as the source of failure. But blame is the song of the anti-teacher.

Lesson 4: Relationships

A teacher can be strict and grade hard, and still have a good relationship with the class. What matters is that the students perceive the teacher as fair and interested in them as individuals. It may sound easy. It's not.

I highly recommend all teachers read the book *How To Win Friends and Influence People* by Dale Carnegie.[1] This book attempts to teach people how to improve their relationships with their friends, co-workers, and spouses. It does this by showing the reader the mistakes and successes of famous people. The book is fascinating; I have read it several times just for the entertainment value of the stories. Carnegie was able to distill into a few basic rules the Do's and Don't's of building successful relationships. The main ideas are simple: First, be complimentary and appreciative; second, almost never be critical; and third, if you have to make people do things for you, try to make them want to do those things. Those rules struck me as solid ideas, and I tried to apply them. But his plan failed for me because I didn't appreciate the better qualities of others. Carnegie had even warned the reader his plan would fail if not applied in a sincere way. But when I was young, I was primarily focused on myself and thought of others only in terms of what they could do for me. Well, who would want a relationship with somebody like that? My friends were pretty much the same, and all we had in common was our association in class and in sports.

Eventually I grew up and gained some social sophistication just by virtue of my age. Today I understand his message. The point of this digression is that maintaining a good relationship with a class has much in common with maintaining a good relationship with a friend or fellow worker. The psychological principles of human interaction are the same. But maintaining a good relationship is not an option for some people. Appreciation of others requires a social sensitivity that is not universal. Some people would like to get along and please others, but the friction caused by compromise and persistent efforts to understand others are too heavy to bear. For these people, the methods and philosophy presented here will not work. They are better off maintaining the traditional teacher-student roles.

I once had an intelligent student named Marco who could pass my tests in computer science, but who wouldn't turn in much homework. Early in the quarter his grade had dropped. I discussed this with him, and he promised to do better. But two weeks later he had turned in nothing more. Here is the conversation we had one day:

> "Marco, you have not been turning in your work for some time. Why not?"
>
> "I lost my disk. I have redone some of the programs, but they are at home."
>
> "But you haven't turned in the written work either."

---

[1] I like this book so much that I offer it as extra credit to all my students. I count it equal to six test questions or about a 50% increase on a test.

"I do them, but I lose my papers."

"Marco, you've got to do better. We've talked before, and it hasn't seemed to help. Right now your grade is a D. I don't want to give you a D, and you don't want to get such a low grade, but what can we do?"

"I dunno."

"Why not make a special effort to turn in some work this week? Reorganize your notebook so that you don't lose so many papers. Spend a little extra time on those partially finished programs and get them to me this week."

"Okay."

"Well, I'm looking forward to giving you at least a C for this quarter, Marco. Don't let me down."

"Okay. Thanks, Mr. Stueben."

How ordinary this student-teacher exchange must appear to an outsider. But it isn't. What is significant here is what is missing: no anger from the teacher and no accusations, beyond the simple facts. Although I made some simple suggestions, I did not lay out a plan or require Marco to develop a plan to improve his grade. He needed to do that for himself. What is present in this conversation is a plain discussion of the facts, a little motivation, and a demonstration to the student that I am on his side. If he receives a poor grade, then I very much want him to feel he has disappointed me.

Shortly after the beginning of the school year, I ask each student to pass in his or her assignment. Many students have not completed them, and some do not even bring them to class. Of course, I know this will happen, because this happens at the beginning of every year. As the students work in pairs, I go around the room and discuss each student's assignment personally with him or her. I say the following carefully rehearsed speech: "I know you want to do well in this class, so it is important that you complete each assignment before you come to class. I'm going to let you finish this assignment tonight and turn it in tomorrow, because I want you to keep a good grade in the class. But in the future, won't you try to complete them before coming to class?" I try to get eye-to-eye contact, but many of them are ashamed and just look down and say something like, "Uh . . . okay. I'll do better." This type of teacher trick is of immense importance in motivating the average student.

Lesson 5: The School Experience

Albert Einstein was one of the few scientists who was also a respected philosopher. His ideas and opinions are still in print and are still influential. The following paragraphs are from a speech he made to educators in 1936 when he was 57:

> Sometimes one sees in the school simply the instrument for transferring a certain maximum quantity of knowledge to the growing generation. But that is not right. Knowledge is dead; the school, however, serves the living. It should develop in the young individuals those qualities and capabilities which are of value for the welfare of the commonwealth. But that does not mean that individuality should be destroyed and the individual become a mere tool of the community, like a bee or an ant. . . .
>
> The most important method of education accordingly always has consisted of that in which the pupil was urged to actual performance. . . .
>
> To me the worst thing seems to be for a school principally to work with methods of fear, force and artificial authority. Such treatment destroys the sound sentiments, the sincerity and the self-confidence of the pupil. It produces the submissive subject. . . . It is comparatively simple to keep the school free from this worst of all evils. . . .
>
> The second-named motive, ambition or, in milder terms, the aiming at recognition and consideration, lies firmly fixed in human nature. . . . Desire for approval and recognition is a healthy motive; but the desire to be acknowledged as better, stronger or more intelligent than a fellow being or fellow scholar easily leads to an excessively egoistic psychological adjustment, which may become injurious for the individual and for the community. Therefore the school and the teacher must guard against employing the easy method of creating individual ambition, in order to induce the pupils to diligent work. . . .
>
> The most important motive for work in the school and in life is the pleasure in work, pleasure in its result and the knowledge of the value of the result to the community. . . .
>
> The awakening of these productive psychological powers is certainly less easy than that practice of force or the awakening of individual ambition but is the more valuable for it. The point is to develop the childlike inclination for play and the childlike desire for recognition and to guide the child over to important fields for society. . . .
>
> Such a school demands from the teacher that he be a kind of artist in his province. What can be done that this spirit be gained in the school? . . . First, teachers should grow up in such schools.

Second, the teacher should be given extensive liberty in the selection of the material to be taught and the methods of teaching employed by him. For it is true also of him that pleasure in the shaping of his work is killed by force and exterior pressure. ...

If you have followed attentively my meditations up to this point, you will probably wonder about one thing. I have spoken fully about in what spirit youth should be instructed. But I have said nothing yet about the choice of subjects for instruction, nor about the method of teaching. Should language predominate or technical education in science?

To this I answer: In my opinion all this is of secondary importance. If a young man has trained his muscles and physical endurance by gymnastics and walking, he will later be fitted for every physical work. This is also analogous to the training of the mind and the exercising of the mental and manual skill. Thus the wit was not wrong who defined education in this way: "Education is that which remains, if one has forgotten everything he learned in school."[2] —Albert Einstein, *Out of My Later Years* (Citadel (Carol), 1984), pages 32–36.

What Einstein does not tell us is how to accomplish the ends. How do we transfer the spirit of independent learning to our students? How do we motivate them to do their best? How do we train them to think analytically? To answer these questions, let me describe to you a typical high school mathematics lesson.

Every year Mr. X teaches the quadratic formula to high school freshmen. He introduces it. He shows them how to say the formula so that it rhymes: *From square of b/ Subtract 4ac/ Square root extract/ And b subtract/ Divide by 2a/ And you're done, hooray!* He discusses the etymology of the word: *quad* refers to a *square* term. After he shows them two examples, he asks the students to solve a problem. He walks around the room looking at the answers. Using the principles of psychology, he praises the students who succeed and encourages the students who fail, but does not criticize. After giving the class another problem, he places some worked examples on the overhead projector and asks the class to find the errors. Finally, he assigns the lesson and sits down to grade the papers collected earlier in the period.

In the sense of Einstein, lessons like this may seem dead: The students are taught how to turn their minds into calculating machines, and even the knowledge they learn is mostly useless, because very few human beings work

---

2 "Education is what you have left over after you have forgotten everything you have learned." —Anonymous, Bartlett's *Familiar Quotations,* 14th ed.

with the quadratic formula. So what has the teacher taught them? What value has the experience been to the students? Here is my opinion: In the process of solving a quadratic formula, many little problems arise—multiplying signed numbers, extracting square factors, canceling and reducing fractions. These problems are inconsequential to a strong mathematics student, but they present difficulty to almost everyone else. For most students to be successful, they must practice, reflect on the solution process, discover their errors, and remedy them. They must learn to live with boring assignments and yet meet their student responsibilities. Such an experience is ideal for a high school student.

What I eventually came to realize is that Einstein's goals were modest goals and could be accomplished with the everyday lesson in the average class. It is not the content of the lessons, but the experience of attempting to master challenging lessons that is of value in life.

Around the turn of the century, a novelist named William Locke was popular. Locke was a former school teacher who had a strong opinion about the teaching of mathematics in school. Here is his assessment:

I earned my living at school slavery, teaching to children the most useless, the most disastrous, the most soul-cramping branch of knowledge wherewith pedagogues with their insensate folly have crippled the minds and blasted the lives of thousands of their fellow creatures—elementary mathematics [specifically advanced high school mathematics]. There is no more reason for any human being on God's earth to be acquainted with the binomial theorem or the solution of triangles, unless he is a professional scientist—when he can begin to specialize in mathematics at the same age as the lawyer begins to specialize in law or the surgeon in anatomy—than for him to be an expert in Choctaw, the Cabala, or the Book of Mormon. I look back with feelings of shame and degradation to the days when, for a crust of bread, I prostituted my intelligence to wasting the precious hours of impressionable childhood, which could have been filled with so many beautiful things, over this utterly futile and inhuman subject. It trains the mind—it teaches boys to think, they say. It doesn't. In reality it is a cut-and-dried subject, easy to fit into a school curriculum. Its sacrosanctity saves educationalists an enormous amount of trouble, and its chief use is to enable mindless young men from the universities to make a dishonest living by teaching it to others, who in their turn may teach it to a future generation. —William John Locke, *The Morals of Marcus Ordeyne* (New York: John Lane Co., 1906), pages 244–245.

Whew! Maybe we should let the air clear before continuing. Some of what Locke says is true, but some is not. There is a popular idea that the study of mathematics and the application of formal logic will increase one's ability to think logically in life, and consequently the student trained in logic will be more successful in solving problems. This is absolutely not true. Most problems that occur in life are solved by examining the causes and formulating a plan based on them: exploration, perception, appreciation, and imagination. The study of mathematics in high school is the recognition of certain common situations and the application of well-known remedies. Consequently, experience with the Pythagorean Theorem, truth tables, and the application of the quadratic formula will be of no help in social and business situations. What will be of help in the lives of our students is self-discipline, responsible attitudes, persistence, love of learning, respect for others, honest self-analysis, and the self-esteem that comes from meeting rigorous challenges. The value of school is not so much in the studies, but through the studies. Contrary to Locke's opinion, there is nothing wrong with teaching algebra and geometry to children, if only the algebra and geometry are not the goal, but rather the medium through which character is developed. Without this "if," Locke's complaint is valid.

Lesson 6: Lectures

I know, indeed, and can conceive of no pursuit so antagonistic to the cultivation of the oratorical faculty ... as the study of Mathematics. An eloquent mathematician must, from the nature of things, ever remain as rare a phenomenon as a talking fish.[3] —J.J. Sylvester, Johns Hopkins University Address (1877).

Consider these two definitions of the word rhombus:

Definition 1 (mathematical). A rhombus is an equilateral quadrilateral. The plural of rhombus is rhombi.

Definition 2 (didactic). A rhombus is a four-sided plane figure with all four sides being equal. Hence, a square is a rhombus. But that is a special case. Usually a rhombus looks like a pushed-over square. (Draw picture.) Now you tell me: Must a rectangle be a rhombus? (No.) Can a rectangle be a rhombus? (Yes.) When? (When it is a square.) Is a rhombus always a parallelogram? (Yes.)

---

3 Found in Henrietta O. Midonick's *The Treasury of Mathematics* (Philosophical Library, 1965), page 768.

The plural of rhombus is rhombi. That is the term we will use in math class, even though rhombuses is in the dictionary. Years ago, my father told me a joke about a zoo keeper who was ordering two mongooses from India. He wasn't sure if the plural was mongooses or mongeese, or even bigoose or polygoose. So he wrote, "Dear Sirs: Please send me a mongoose. Oh, by the way, send me another one, too." What in the world is a mongoose, anyway? (Short discussion follows.)

But back to our definition. Is a parallelogram always a rhombus? (No.) Is a parallelogram sometimes a rhombus? (Yes.) Can a pushed-over rectangle be a rhombus? (Yes.) When? (When it is a pushed-over square.)

Both definitions contain essentially the same information. The first one is absolutely adequate for anyone who has a knack for mathematics. Such a person can figure out the relationship between squares, rectangles, and rhombi by themselves. But most people are not like that. That is why the second definition-explanation (actually a mini-lesson) is better for a high school class. And even that is not enough. The students will later need drill on the definitions and little games involving the words.

The need for all this extra teaching is not due to a lack of intelligence or aptitude. High school definitions are usually simple. The problem is that to learn mathematics requires interest, and most people just can't get interested in the subject without help. If the teacher makes it interesting, as with the mini-explanation and the mongoose story, then more progress can be made.

Many educators have railed against the lecture method.[4] I think the reason for this is that many teachers do not lecture well. And if a teacher is a poor lecturer, then the educators are right, and the lecture method should be used sparingly. But if the lecture is done well, then it is one of the best ways to teach. Here is the trick: Give interactive lectures, not passive lectures. For example, before completing the statement of a theorem, ask the class to predict what you will write. Ask the class to guess the next step in a proof. Ask them negative questions—e.g., find counterexamples and errors. Negative questions are better than positive questions, but they are harder to formulate. I often stop and ask my classes to do a short numerical problem, then I walk around and confirm the correct answers. Generally, I do not go more than three minutes without obtaining a response.

---

[4] "Denunciations of examinations, like denunciations of lectures, is very popular now among educational reformers and I wish to say at once that most of what they say, on one topic and on the other, appears to me to be little better than nonsense; and it has always seemed to me that mathematics among all subjects is, up to a point, the subject most obviously adapted to teaching by lecture and test by examination." —G. H. Hardy, *Mathematical Gazette* (March 1926).

The best method for formal lecturing of theorems is described by the famous mathematician Paul Halmos in his autobiography:

> For each theorem, describe the context it belongs to, the history that produced it, and the logical and psychological motivation that makes it interesting; state it, perhaps roughly, intuitively at first, and then precisely; and finally use it—show what other results it makes contact with, and imagine what follows from it. ... Proofs can be looked up. Contexts, histories, motivations, and applications are harder to find—that's what teachers are really for.
> —Paul Halmos, *I Want To Be a Mathematician* (Springer-Verlag, 1985), page 264.

Where are we to find these contexts, histories, and motivations? The answer is by studying our subject throughout our lives, by reading history books and biographies, by attending talks and conferences, by discussions with our colleagues, by skimming textbooks, and by being on the lookout for patterns and relationships. Cut out Halmos' statement and tape it to the front of your textbook. Try to place some of these attributes in each of your lectures.[5]

Lesson 7: Who Has Intelligence?

Every year I teach proofs in geometry. Some students learn easily, and others struggle. I like to tell them a story about talent in problem solving. But before I tell it to them, I give them this puzzle.

> A few years after I had graduated from college, I moved to Northern Virginia and became a school teacher. One Christmas I flew into Midway Airport in Chicago and approached the car rental agency that was supposed to have a reserved car for me. They didn't have the car. Chicago had been hit with an extreme cold wave, and they couldn't get most of their cars to start. The other car rental agencies had the same problem, and on that day all the rental cars in Chicago that would start were rented out. But I needed a car immediately to drive to McHenry, a small town 50 miles away where my parents lived. What was I to do?

---

[5] Here is another piece of advice I liked: "Much, much more has been written about good teaching, but all the advice I have seen can, I think, be boiled down to about this: The good teacher is a human and mature person who knows his subject thoroughly, has a keen interest in it, and tries to get it across to his students in a thought-provoking fashion." —Roy Dubisch, *The Teaching of Mathematics* (John Wiley, 1963), page 4.

I let my students suggest solutions for a minute, and then I tell them the following story.

> In the early 1970s when I was a college student, I dated a girl named Diane. One day when I was at her apartment, the lights went out. There was a power failure. I suggested that if she had any candles, then we could light them and sit on the floor, and it would be really romantic. She said that she had candles, but that she didn't have any way to light them. "Maybe we could light them off the stove," I said. No, the stove was electric, not gas. So I thought about the problem, and there didn't seem to be a solution. Then Diane thought about it. She suggested that we go out to the car and try to light them from the cigarette lighter. I never would have thought of using the car's lighter. That was the first time I realized how smart Diane was. As the years went by, I saw her solve all kinds of problems that would have defeated me. Eventually she became a manager in an important Washington, D.C., law firm. She became well paid for her problem-solving ability. Whether the problems were research or morale, she seemed to find a solution.
>
> The reason I'm telling you a story about how smart this girl was is that there are three things she couldn't do. The first is proof in geometry. She told me that she memorized her way through her high school course, and that it was an unpleasant struggle. The second thing she couldn't do was play chess. I tried to teach her, and although she could memorize how the pieces moved, that was the limit of her understanding of the game. The third thing she couldn't do was write a computer program. I was taking my first course in programming when I met her. I tried to teach her the FOR-loop. She could memorize the syntax, but she had difficulty applying it to simple accumulation problems.

I discuss Diane with both my geometry and computer science students to show them that being a weak mathematics student or computer science student is not a reflection on their intelligence, their value to society, and their importance as people. Conversely, the ability to grasp mathematics quickly is not much of an indication of any intelligence other than the ability to understand mathematics easily. That is the message. After I finish the talk, I ask if anyone has thought of a solution to the rent-a-car problem. I have only once had a student solve this problem in all the years that I offered it. So here is the rest of the story.

> I didn't have a clue about how to solve this problem, yet I wasn't too worried about finding a solution, because I had a secret weapon:

I was traveling with Diane. First, we made several calls to other rent-a-car agencies in the Chicago area and confirmed what the local agency had told us: There were no available cars to rent in the area. Then it was time for Diane to do her stuff. It took about thirty seconds for her to say: "Well, if we can't rent a car, then let's rent a truck." Why couldn't I have thought of that? We learned the address of the airport from a worker and called a truck rental agency nearby. Evidently there were all kinds of trucks available, because the first place we called had several. In fact—and this is almost unbelievable—they even had a car to rent. There must have been scores—if not hundreds—of travelers in Chicago that day who desperately needed personal transportation. The transportation was within their reach in the form of small trucks and a few cars, but they just didn't have the intelligence of Diane.

Experiences like this have given me a different perspective about teaching mathematics which I didn't have as a beginning teacher. I now give a high priority to making students aware that intelligence comes in many forms. I make a point of complimenting clever thinking even on matters trivial and unrelated to the course material.

Lesson 8: The Scholar

The scholar or ambitious student is the opposite of most classroom students. The scholar searches for knowledge; the average student first searches for a grade, and then pleasure in understanding. The scholar is self-motivated, disciplined, and mature about the difficulties of learning. The average student lacks the talent, lacks the discipline, and is immature and foolish about scholarly things. The average student is easily distracted by the problems and temptations of life. Most students do not see the big picture in either their courses or their responsibilities in school.

Scholars thrive in a competitive environment. They respect intellectual power and have contempt for stupidity. Guess who will become the teachers for the next generation? The scholars of the last generation. The problem is that the scholar's attention to book knowledge can too easily become a hazard to effective teaching. Why? Because the scholar is in love with his subject, and the subject is secondary to a student's classroom experience.

Lesson 9: The Lazy Student

The most common complaint uttered by teachers about students is that too many of them are lazy. But look into the mind of a student who you think is lazy. You'll find an incredible morass of misconceptions, ego problems, irresistible temptations, unrealistic expectations, bad habits, denial of reality, refusal to face problems, dependency on fantasy, history of abuse from others, and more. We should be thankful we are not like that. I believe there is no such thing as a lazy student. What is seen as lazy is distraction and lack of motivation. The difference between these terms is that "lazy" connotes unworthiness, and "unmotivated and distracted" connote a handicap.

Think back on your life about some of your regrets. If you are like most people, you haven't killed anybody or destroyed a career. But you might have been rude and unappreciative of another person. You may have cheated on a test, or stolen something as a child. Whatever it is, why did you do it? The answer is that you were carried away by emotion you couldn't control, or that you lacked the maturity to think beyond your immediate desires, or that you didn't have all the information you needed, or that you judged too quickly, or you thought you were right and didn't re-assess, or a score of other reasons. If someone else were in your body with the same feelings, the same experiences, the same perspective and ideas, then it is hard to believe they would have acted differently. Although life generally makes us face the consequences of our actions, I am saying that we are not responsible for our mistakes. We don't intentionally make mistakes; we make them because we are blind to the consequences. You and I do the best we can with what we have. And so it is with our students. The idea of a lazy student is the invention of the Devil.

Lesson 10: The Dishonest Student

In 1651, the English political philosopher Thomas Hobbes wrote a book called the *Leviathan*. His book, which is still in print, is a study of the human psychology of politics. Read what he has to say about telling the truth:

> The doctrine of Right and Wrong, is perpetually disputed, both by
> the Pen and the Sword: Whereas the doctrine of Lines, and Figures,
> is not so; because men care not, in that subject what be truth, as a
> thing that crosses no mans ambition, profit, or lust. For I doubt not,
> but if it had been a thing contrary to any mans right of dominion, or
> to the interest of men that have dominion, That the three angles of
> a Triangle should be equall to two Angles of a Square; that doctrine
> should have been, if not disputed, yet by the burning of all the books

of geometry, surpressed, as farre as he whom it concerned was able.
—Thomas Hobbes, *Leviathan* (1651; Penguin edition, 1985), part
I, chapter 11, page 166.

People will justify doing just about anything they want. We know Hitler
tried to justify invading Poland, and millions of Germans believed him. Our
desires for survival, success, and comfort are so powerful that they trick us
into believing that what seems to be to our advantage is actually the right
thing to do.

I would like to tell you about two great men. The first is Jesse Stuart,
a poor farm boy from the Kentucky hills who grew up to be a great teacher
and writer. His first teaching assignment after graduating from college was
in a rural school house with 14 high school students. The entire faculty consisted of just one teacher: Jesse Stuart. What could one expect from such a
situation? What Stuart found was 14 students, each with a strong desire for
more education. Within two weeks they had each read all the books he had
brought to read for himself. He got them other books, but they read those too.
He was so impressed with their desire to learn that he risked his life twice to
bring them more books. The school was in a valley that had only one road
in and out in the winter. He could walk it or go by mule train. One day he
made the mistake of walking and became lost in a snowstorm. Then he fell
into water and got both feet wet. Only by the good luck of wandering into
stacks of fodder in a farmer's field did he survive. He made a tiny hut from
the fodder shocks and put on all the clothes he was carrying in his suit case.
In the morning he discovered his shoes had frozen in the night and would not
fit on his feet. So he wrapped shirts around his feet and put on his galoshes
and continued his trip.

On the way back to the valley Stuart had overlooked the fact that books
are very heavy. By the time he returned to his valley, he was exhausted and
had no feeling in his feet and hands. Only by the kind and intelligent help of
a neighbor did he avoid frostbite.

That was the kind of a person Jesse Stuart was all of his life. For years,
his book *The Thread that Runs So True* was required reading in college
education courses. When he was 49 and recovering from a heart attack, his old
high school became desperate for a principal. Stuart was financially solvent
through the writing of his books. He didn't need the money, but he took the
job even though his doctor wouldn't allow him to climb stairs. Why am I
relating the life of Jesse Stuart to you? Because although Jesse Stuart was a
great man, he was also a school cheat. I'll let you read this in his own words.

I couldn't tell Superintendent Larry Anderson I had for my room-
mate at Lincoln Memorial, Mason Dorsey Gardner, who had won

the coveted award of twenty-five dollars for being the best math student. I couldn't tell him Gardner had worked my algebra and I had written his themes. —Jesse Stuart, *The Thread That Runs So True* (Scribner's, 1949).

Stuart was old enough to know right from wrong, yet I detect no remorse in his comments. So what are we to think? Before I give you my opinion, I want to tell you about another great man.

J. L. Kelley taught mathematics at the University of Notre Dame, the University of Chicago, and the University of California, Berkeley. In 1947 he refused to sign the loyalty oath imposed by the University Regents and was dismissed from his tenured position although he had a wife and three children to support. Kelley never was a Communist, but he felt it vitally important that he shouldn't have to declare himself a non-Communist. He felt loyalty oaths were morally wrong and wouldn't sign one.[6] Years later when the oath was declared unconstitutional, he was re-hired at Berkeley and served two terms as department chairman. He also wrote a famous topology textbook and set up a math-for-teachers major. I was very impressed when I read this about Kelley. I was also impressed by this passage from his recollections:

I undertook three courses in education in my first three semesters at UCLA in order to prepare for a secondary credential. The courses were pretty bad and besides, the grading was unfair, e.g., I wrote a term paper for Philosophy of Education and got a B on it; my friend Wes Hicks, whose handwriting was better than mine, copied the paper the next term and got a B+, and our friend Dick Gorman typed the paper the following term and got an A.—John L. Kelley, Peter Duren, ed. Vol. 3, *A Century of Mathematics in America* (American Mathematical Society, 1989), pages 474–475.

Now why would a highly principled person like Kelley give his term paper to other students to copy? Was it youthful indiscretion? Probably not, because he wrote these sentences in 1987, two years after he retired from teaching and mentions his cheating only in reference to the grading system. I suspect that Kelley thought then and still thinks the course was a waste of time. I'm sure he felt he was helping his friends by letting them copy his

---

[6] "One day [c. 1949–1951] when I visited Berkeley, I met a member of the faculty who was a very ardent communist. He wanted to join the party, but the party whouldn't let him so that he could swear truthfully that he was not a card-carrying communist. I asked him where the party stood on the loyalty oath. He replied that the instructions were to sign and to urge everybody else not to." —Richard Bellman, *Eye of the Hurricane* (Singapore: World Scientific, 1984), pages 155, 156.

work. Jesse Stuart probably chose to cheat because of pride. He wanted to get a good grade in a course that was difficult for him.

Our students often think our courses are a waste of time. They have difficulties with certain classes, with their teacher's attitude, with their classmates, with their parents, and with their own failings. They are immature, unsophisticated, and irresponsible. In the midst of all of this, they begin to cheat. On a surface level we can see why they do it: It saves time, it gives them a better grade, it avoids work. On a deeper level, it goes against their moral code. So they fool themselves into believing their cheating is justified by special circumstances. Cheating by a student is not indicative of low morals, weak character, a bad upbringing, or a disreputable future. Kelley and Stuart prove that point. Cheating is usually a mark of insecurity, immaturity, or fear. Consequently, strong punishment is usually out of order for cheats. My suggestion is to give them a zero on the assignment, or make them do it again and require them to write a one-paragraph essay explaining their actions. I try not to involve the administration or parents. If the student is hopeless, I try not to become angry, or to "teach him a lesson in life." I say my piece, and then we go our separate ways. There is a limit beyond which a teacher can be effective. In any case, the most effective teacher of honesty and respect is not our words, but our actions in the classroom.

By the way, are you curious about how Jesse Stuart treated cheaters in his own classes? Here is a reference I found in one of his writings:

> In order to get things going, I tell students to write about anything they like and put it in any form they please. The race begins. They envy each other. They try to do better work than their friends, and they jump so far ahead that I find some students start copying. They don't intend to be outdone. I have seen teachers who would throw a student out of the class for a trick like this. But why should the student be thrown out? You understand he has pride, and hates to let another surpass him when he stands up and reads before the class once a week. —Jesse Stuart, *To Teach, To Love* (World, 1970).

Lesson 11: On Being Personable

I try to memorize the correct pronunciation of my students' first and last names, and I use their names when I speak to them. When a student approaches me, I speak first and use the student's name. When something unusual has happened to a student, I ask him or her about it in class and let the student talk about it. When a student gets an A on a test, I write a big red three-inch capital A on the paper and make a comment like "Excellent work,

Maria!" I look for opportunities to give sincere compliments. If a student comes to class particularly dressed up, then I give him or her a quick compliment. If a student wears an interesting T-shirt, then I ask where he got it, and maybe if he believes the message on the shirt. At quick moments before class, after class, or in conversations during group work, I ask my students questions about subjects I think they will enjoy answering.

All of this may sound like cutesy education stuff unrelated to real teaching, but it isn't. It is serious teaching. This kind of behavior on the part of the teacher creates a learning environment that is less intimidating, less fearful, and less impersonal.

Whenever a student comes to me with an unusual request, such as being allowed to re-take a test, or to be allowed to turn in late work, I listen to their reasons and then help them with their argument. Typically a conversation runs like this:

> "Mr. Stueben, I had a game last night and didn't have time to study for the test. Can I take it later?"
>
> "So you feel that my test will be unfair to you because your football practice interfered with your study time. And you couldn't let your teammates down by not coming to practice."
>
> "Uh . . . Yes, Mr. Stueben, that's right."
>
> "Well, you know, Johnny, modern educational theory would agree with you. Grades are supposed to reflect a student's knowledge, not give a false reading because the student didn't have an opportunity to study for a test. How about if you take the test now as a practice, but I won't count your grade. Then make an appointment with me for a re-test next week. I'll get a copy of a test from another teacher."
>
> "Thanks, Mr. Stueben."

Of course, what I have difficulty not saying is this:

> "You must be joking! When you went out for football, you were told that it was your responsibility to meet all assignments. I'm sure your coach would have given you one day off from practice if you had asked him. Whenever there is a conflict between sports and schoolwork, schoolwork must come first. You should learn to be more responsible. Blah, blah, blah."

Actually, the sports dynamic is so strong that most young people can't resist it. So the idea that school work is expected to come first for an athlete is unrealistic. Second, teenagers are even more immature about responsibilities than adults. Most students will see themselves as victims in situations like this, even though the trouble is mostly their own doing. Third, to keep order and to maintain a workable relationship with students means the teacher must be perceived as fair. But to be seen as being fair means the teacher must actually be more than fair in some cases. It is the inflexible and socially immature teacher who trots out his or her high principles and refuses to bend.

Sometimes I need to talk seriously to a student about his or her behavior. I like to ask the student to step out of class and walk down the hall with me. When we get to the end, we walk back. When you walk with somebody, you don't have to look at them when they talk. This helps students better concentrate during a tense conversation. My typical talk goes something like this:

> "Alice, I really like you, and you seem to be a really neat person. But we are having too many unproductive conversations, and they are affecting our relationship. If we don't stop this, we will come to dislike each other. When I ask you to do something for me in class, I'm not trying to pick on you—I really need you to do it for me. Don't you think we should both try to get along better with each other?"

Sometimes it works, and sometimes it doesn't. But a student is more likely to react positively to my request for a behavior change if he or she believes I am sincerely interested in all my students. And this impression is fostered by my listening to students and drawing them out in short conversations about what interests them.

What I haven't explained here is how to approach teenagers. How to be sympathetic, but not patronizing; how to give in without being taken advantage of; how to give a compliment without seeming insincere; how to guide without criticizing. Nobody can explain how to do these things, because this is the stuff of life, this is human interaction on the most basic level.

Lesson 12: The Anti-Teacher

What I am going to say next is pretty strong. I believe I have a right to preach a little because I have been guilty of all these sins myself. In an attempt to deal with the frustrations of teaching, it is possible to go the wrong way and become what I call an anti-teacher.

The anti-teacher is a person who tries to control others and to bend them to his way of doing things. He is a bully. The anti-teacher is a name-caller, and his favorite words for students are lazy, stupid, slacker, dishonest, and troublemaker. The anti-teacher treats students in a cavalier manner, because he is serious and they are silly. The anti-teacher wants to build a better world,

The Teacher	The Anti-Teacher
Concerned	Ambivalent
Helping	Ignoring
Respecting	Contemptuous
Understanding	Sarcastic
Friendly	Aloof
Rewarding	Punishing
Listening	Preaching
Empathetic	Critical
Humorous	Angry
Forgiving	Blaming
Motivating	Insisting

FIGURE 1

but becomes frustrated when he sees that his efforts to persuade other people fail. Figure 1 is a chart I made to contrast the teacher with the anti-teacher.

Becoming an anti-teacher is the worst thing imaginable that can happen to a teacher. In attempting to make a better world, he or she makes things worse. Anti-teachers usually can't be talked out of their philosophy, because they are sure they are right. A statement of the famous psychologist Alfred Adler comes to mind here: "It is the individual who is not interested in his fellow human beings who has the greatest difficulties in life and causes the greatest injury to others."[7]

Conclusion

In 1989 I had a brief conversation with an algebra teacher who was getting ready to retire. Just to make idle conversation, I asked: "After you retire, are you going to write a book about teaching algebra so that the rest of us can learn your tricks?" He replied, "No, I don't think I ever quite got the hang of teaching the subject." The moment he said that, I realized that after 14 years of teaching mathematics and computer science, I had not gotten the hang of teaching these subjects either. Earlier in my career I had worked hard to master my craft and had reached a certain level of competence. I could control my classes and get the lessons across, but I had not progressed beyond

---

[7] Alfred Adler, *What Life Could Mean to You* (1931; retranslation by Colin Brett, Oxford, England: Oneworld Publications, 1994), page 210.

that point. I was a minimally competent teacher. It took me 14 years to realize my errors, and another five years to find my way. I have written this chapter in the hope that others might rethink their roles as teachers and will not later look back at their careers as nearly 20 years wasted before the blackboard.

An Example from my Teaching

The first two tests in my 1993 geometry classes were both machine-gradable and were both prepared prior to the opening of school. In fact, they were the previous year's tests. This gave me more time to do the extra work required at the beginning of the year. The first test was a little tricky. The students did not do well. After I returned the tests, I had them do a post-test analysis that consisted in asking two questions about every problem they got wrong: a) What is the correct answer, and b) Why did you get this problem wrong— a question of psychology, not mathematics. I told them that to succeed they needed to study their own minds as well as the course material. I also explained why so many of them had a poor performance: They were new to high school; they were not used to my tests; geometry was a difficult subject for a lot of students; and they needed to adjust. I told them that if they did better on the next test, then I would count that grade twice and throw out the first grade, but they would be tested on the old material as well as the new material.

Shortly before I tested again, I reviewed all those topics that would appear on the second test. The first test (chapters 1–3 in Moise-Downs's *Geometry*) had been on various topics. The second test occurred after chapter four (proofs) and required a short six-point proof at the end.

The grades were better, but some students fell down on the proof. So I announced that anyone could retake the proof part of the test by coming to my room during lunch or after school on Wednesday.

Four students showed up after school: Fatima, a B student who misunderstood about placing the GIVEN parts in a proof; Heather, an extremely conscientious student who wanted to raise her grade from 5/6 to 6/6; Tim, who scored 0% on the proof; and Nikoli, who also scored 0% on the proof.

Fatima and Heather turned in make-up proofs that were almost correct. By asking them a few simple questions during the grading process, they were led to see their errors and to make corrections during the grading process. They each received 6/6.

Nikoli turned in a wretched proof. I pointed out the first error and he corrected it, then I proceeded from error to error, letting him correct each step. This took some time. He often had to return to his seat to think about his proof for a few minutes before he could suggest the correction to the next step, and his corrections were often wrong. Eventually he completed the proof. While he was doing this, I kept the mood lively by complimenting him on his little successes. When at long last he completed the proof, I told him I wanted him to do only one more thing. I took his paper and cut off all of his work. Then I taped the remainder to the top of a blank page and asked him to redo the entire proof. He did it in about ten minutes making only two errors: He gave the wrong reasons for two legitimate steps. During this time, he looked like he was in agony trying to recall what we had previously done. I helped him correct the last two errors, then marked the top of the proof with a big 100% and complimented him for working so hard to master proofs and to improve his grade.

Tim, the fourth student, seemed a little better than Nikoli. But I had run out of time when I finally guided him through the errors. So I cut off the proof part and taped the remainder to another page and told him to work on it at home. I said he could consult any book he wanted, because we had worked through the proof without a book, but he was on his honor not to seek help from another person. I scored all four 100% in my grade book before I left and hoped I didn't forget to check Tim's proof the next day.

What was gained by all this hand-holding or spoon-feeding, as some people would call it? By giving a re-test, the class became convinced that I was concerned with their grades as well as their knowledge. They came to trust me as their leader. By working with them after school, two weak students came to believe there was hope—at least with me as their teacher. These kinds of tactics lead students to do their best.

But it is easy to misapply a good idea. My remarks to each of the four students *necessarily* had to be encouraging, friendly, and patient. When I encountered an error in Heather's paper, I said, "That's wrong, but if you can tell me the correct answer now, then I'll count it right." The four encounters became games in which both teacher and student teamed up to maximize the grade. If any of the four students had come away with no appreciable grade change, or detected insincerity in my remarks, or felt that helping them was not enjoyable for me, then the real benefits of these encounters would have been lost.

The drawback was that it took an hour of time I needed to prepare for the next day's lessons. It is impossible to offer this kind of help on a regular basis. Exactly what tactics are to be used with which classes will vary among teachers. I have no argument with a teacher who claims students should almost never be allowed to re-take tests. But then some other method must be found to raise the morale of the weaker students and to maintain the class's confidence in the teacher.

2
Humor in the Classroom

My all-time best teaching trick was to introduce a "curiosity-of-the-day" into my basic algebra and regular geometry classes. Just before I was to give a lesson I would tell them something curious about the world or tell them about some problem I solved in my life. Adults know many things that high school students don't know. For example, kids don't know about the process of buying a car or a house. They don't know about the tricks salespersons use to get their customers to make larger purchases. They have never gone through a world almanac and looked at the marriage and divorce statistics. These things are fun to discuss with a class and to ask them questions like "why can't people seem to stay married?" For several years, after the second quarter began, I tried to have a curiosity each day except on test days. Occasionally the discussions lasted 20 minutes or longer. Those were great times because the students became so interested in a topic. Some of the students wanted to keep the conversations going because they knew it would make for a shorter math lesson. That didn't bother me, because enthusiastic class discussions are such wonderful learning moments. Years later, when I would meet former students on the street, they would ask me if I was still presenting curiosities each day, and they would tell me how much they enjoyed them.

Eventually I began to teach the more advanced geometry and computer science classes, and I noticed that these students were less interested in curiosities unrelated to the subject. So I stopped giving curiosities unless they related to the daily lessons. This chapter contains some of the subject-specific humor, anecdotes, and curiosities that I still use.

Teaching Understanding

> ## The Great Student Trap
> Teacher: I will try to help you understand.
> Student: Please, sir, would you just let me memorize
> the rules necessary to pass the tests?

A group of academics was traveling by train to an important international conference. In the last car there was one small party of mathematicians and one small party of economists. Each economist had a ticket, but only one of the mathematicians had a ticket.

"How are all of you going to travel to the conference with only one ticket?" asked the leader of the economists.

"Oh, we have a method," replied his mathematical colleague.

Shortly before the conductor entered the car, all the mathematicians went into one of the two toilets in the rear of the car. The conductor soon arrived, collected a ticket from each economist, and then knocked on the toilet door. The door opened slightly, and a hand poked out with a ticket. The conductor collected it and left.

When the mathematicians returned to their seats, the chief economist said, "Ah, very clever. This is yet another mathematical method that we economists can apply."

About a week later, the two parties found themselves in the same train car making the return trip. This time only one of the economists had a ticket, but no mathematician had a ticket.

"How are all of you going to ride back with no tickets?" asked an economist.

"Oh, we have a method," replied a mathematician.

Shortly before the conductor entered the car, all the mathematicians went into one toilet. Then all the economists went into the other toilet. Then one mathematician came out of his toilet and knocked on the door of the other toilet. The door opened slightly, and a hand poked out with a ticket. The mathematician collected it and returned to his toilet.

Moral 1: Do not apply methods that you do not understand.

Moral 2: Beware of shifting domains.[1]

---

[1] Dr. Sergej L. Bezrukov, University of Paderborn, Germany (Usenet correspondence, October 1993).

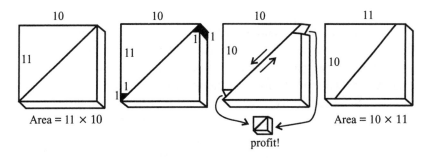

FIGURE 2

Consider the following scheme to make millions. Obtain a slab of gold $10'' \times 11'' \times 1''$. Score it diagonally as shown in Figure 2. Then cut out two little one-inch triangles at the corners. These triangles are the profit. Finally, slide the two slabs together again and note that the rectangle is now $11'' \times 10'' \times 1''$—the same size as it was before we removed the small triangles. By repeating the process, we can produce an unlimited supply of gold![2]

You are richer than you think:

$$1¢ = \$0.01 = (\$0.1)^2 = (10¢)^2 = \$1$$

An excellent book for paradoxes and fallacies that will appeal to high school students is Eugene P. Northrop's *Riddles in Mathematics* (Van Nostrand Reinhold, 1944). Find this masterpiece, read it, and integrate its treasures into your lessons. Here is one of his tricks:

$$3 > 2$$
$$3\log_{10}(1/2) > 2\log_{10}(1/2)$$
$$\log_{10}(1/2)^3 > \log_{10}(1/2)^2$$
$$(1/2)^3 > (1/2)^2$$
$$1/8 > 1/4$$

---

[2] This puzzle is a variation of one found in Martin Gardner's *Mathematics, Magic and Mystery* (Dover, 1956), pages 129–130. His book contains two chapters on the paradoxes of "geometrical vanishes."

Around 1941, while in the military, my father took a test of mental ability. One of the questions was this: If one man can build a house in 30 days, how long will it take two men to build the same house? My father claimed that he didn't know. The psychologist was surprised that a college graduate could not answer such an apparently simple question. My father replied that it would take the same amount of time for the concrete to dry no matter how many men were building the house. And, further, in lifting materials to the roof, two men might be four to ten times as efficient as one man. After some thought, the psychologist agreed and dropped the question. That question was a simple misapplication of direct proportion.

When introducing direct proportion to a class, I take the opportunity to point out its misuse by staring dramatically at the students and saying:

> What we have just seen is called ... (pause) ... the Law of Direct Proportion. It is a very powerful law and has important real-world applications. For example, if one aspirin does me a little good, then I know that 100 aspirin will do me 100 times as good. And if one boat can cross the ocean in ten days, then ten boats can cross the ocean in one day. (long pause) Uh ... is that right?

And then we would discuss why the application of direct proportion sometimes gives a correct answer, and sometimes does not.

Wordplay, Puns, and Other Mathematical Atrocities

Around 1983, I was teaching constructions in geometry and decided to hold a contest between the boys and girls to see which group could best draw circles on the blackboard. The students didn't realize that using a blackboard compass is quite awkward and requires practice. The worst circle was drawn by a boy named Nicholas. Near the end, the girls were ahead by one point, but there was one more girl in the class than there were boys, so I had to pick one boy to come up a second time. On a whim I said: "Okay, the girls can pick the boy who gets to go up twice." Of course, they all shouted: "Pick Nick!" Suddenly, from the back of the room, Cheryl Turner said: "Don't be silly, people, we can't pick Nick; it's too early in the year." Whenever I have a student named Nicholas, I try to pull off this pun. Thanks, Cheryl.[3]

1. There are only two things you need to know in life: First, don't tell people everything you know.
2. There are three kinds of mathematicians: those who can count, and those who can't.[4]

Many students don't understand these jokes at first. But when some students laugh, it motivates others to search for a point, and thinking about subtle relationships is exactly what we want our students to do.

Professor Blackie of Edinburgh, being indisposed one day, caused to be posted on the door of his lecture room the following notice: "Professor Blackie will not meet his classes today." A student who was a bit of a wag erased the "c" in "classes." The professor hearing of it sent a messenger with instructions to erase the "l."[5]

My freshmen geometry students have always appreciated that little story. John Stuart Blackie (1809–1895) was a professor of Greek at Aberdeen and Edinburgh for 30 years. He was also something of an amateur philosopher. Here are two of his epigrams that I found in a 1936 book of quotations:

1. Eccentricity is originality without sense.
2. Absolute rules are a device of cowardice to escape the difficulty of decision when an exceptional case occurs. A consistent refusal is always easier than an occasional compliance.

I especially like the second one, except that "cowardice" is not right—it should be "convenience." Being asked to make an exception to a rule is an almost-daily task for high school teachers. Some teachers are strict, and other teachers give in a lot. So what's the best policy? The best policy must depend on the teacher, the class, and their relationship.

---

3 Another Cherylism: Teacher: "Does this make sense?" Cheryl: "Well yes, I understand it, but no, I doubt that it will make any money for me."

4 Attributed to Enrico Bombieri in Betsy Devine and Joel E. Cohen, *Absolute Zero Gravity* (Simon and Schuster, 1992), page 153.

5 Rev. G. A. Carstensen, found in *Modern Eloquence,* vol. XIV (P. F. Collier & Son, 1914), page 117.

"Nothing is so unequal as the equal treatment of unequals."[6] The trick, of course, is always *to be seen* as treating everyone equally. I can recall two students, in different years, who came to see me after school and asked me to let them retake a test. I agreed and said, "Okay, but do not tell your classmates, because if everyone wants a new test, then the grading will become impossible." Was this fair? In my mind it was, because I gave a re-test to every student who asked for one.

When Isaac Asimov was a high school student (circa 1934 in New York City), his English teacher had his class read the famous poem *Abou Ben Adhem* by James Leigh Hunt (1784–1859). In the poem, a man called Abou is criticized by an angel for his lack of piety. Abou responds that at least he loves his fellow man. The last line of the poem is a surprise. The angel shows Abou a list of those blessed by God: "And lo!, Ben Adhem's name led all the rest." The teacher asked the class why the name of a man so unpious as Abou was placed first. Fourteen-year-old Asimov shouted out: "Alphabetical order!" He was kicked out of class for that. The trick is to keep students under control without repressing their creativity. That witticism was brilliant.

I have my geometry students draw this picture (Figure 3) in their notes, and then I ask them what it means. (The intersection of any two planes is always a straight line.) One day after class a student came up to me with a cartoon drawing of two airplanes colliding in mid-air. "What is this?" I said. "It's what you were talking about in class, Mr. Stueben: the intersection of two planes."

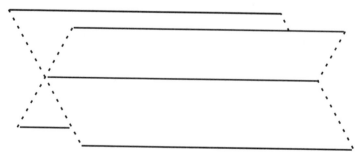

FIGURE 3

---

[6] Floyd G. McCormick, Jr., *The Power of Positive Teaching* (Krieger, 1994), page xix.

Points
Have no parts or joints.
How then can they combine
To form a line?

—J. A. Lindon[7]

This is a cute rhyme to read to a geometry class when they begin to discuss points and lines.

I have yet to see a problem, however complicated, which when you looked at it in just a certain way, did not become still more complicated.[8]

"Never smile until after Christmas."—Old teacher saying. This is good advice for some classes and for some teachers. And for all my classes, I try to avoid telling jokes and stories until after the first quarter. Even after the first quarter, classroom humor is tricky. If a pun relates to the topic being discussed, then students might enjoy it, especially if it is a clever pun, if it is told well, and if the lecture is going well. But classroom humor cannot be used to rescue a poor performance. A joke delivered into a class that is frustrated or bored will backfire.

Logarithms

The best explanation of a logarithm that I know is this: *A logarithm of a number is the number's size relative to some scale (base). In physics the best scale to use is determined by the phenomena being described. In theory it doesn't matter which base you use, because logarithms to different bases are only multiples of each other.*

---

7 "A Clerihew ABC of Mathematics," *Recreational Mathematics Magazine,* no. 14 (January 1964), pages 24–26.

8 Attributed to Poul Anderson (science fiction author).

number	number of zeros	log (log base 10)	ln (log base e)
10	1	1	$1 \times 2.302\ldots$
100	2	2	$2 \times 2.302\ldots$
1000	3	3	$3 \times 2.302\ldots$
10000	4	4	$4 \times 2.302\ldots$

Before the invention of hand calculators in the early 1970s, if engineers wished to multiply two large numbers, and would be satisfied with an approximation, then they would use the formula $\log_n(ab) = \log_n a + \log_n b$ with a book-sized table of logarithms or a slide rule.

> To reproduce, Noah's snakes were unable.
> Seems their fertility was somewhat unstable.
> So he made a bed
> Of tree trunks and said,
> "Even adders[9] can multiply on a log table!"
> —Anonymous.

Mathematical Q and A

Q1: How many psychologists does it take to change a light bulb?
A1: Only one, but the light bulb must really want to change.

Q2: How many mathematicians does it take to change a light bulb?
A2: None. The mathematician merely gives the light bulb to a psychologist, thereby reducing the problem to a previously solved case.[10]

Each year I tell this joke to a class just before I reduce the solution of a problem to a previously solved case. Since mathematics has been taught for over two thousand years, it amazes me that we don't have more jokes like this.

Q: What are the two best-known "ants" in high school mathematics?
A: The answer is *determinant* (not *determinate*) and *discriminant* (not *discriminate*).[11] This is a useful wordplay for poor spellers.

---

[9] An *adder* is a common poisonous snake.

[10] Thanks to my former student Matt Blum (1991).

[11] Note that the frequently misspelled word *existent* is not an "ant" word.

Q: How much is a millihelen?
A: The amount of facial beauty necessary to launch a single ship.[12]

Q: What is the shape of a kiss?
A: Elliptical (e-lip-tickle).

Q: Why does *e* occur so often in mathematics and science?
A: Because *e* is the most frequently occurring letter in English writing.

Q: What is brown, wrinkled, and says less is more?
A: Occam's raisin.

Occam's Razor, a concept of the English philosopher William of Occam (1300–1349), was later paraphrased as "entities must not be unnecessarily multiplied"—i.e., simplest is best. This is the name for the tendency to choose the simplest hypothesis that explains a set of facts. The other hypotheses are cut away. The following quotation is by Freeman Dyson, a professor of physics at the Institute for Advanced Study in Princeton:

> Dirac[13] maintained that the way to make discoveries in physics was to look for something beautiful, but I think it led him very badly astray. First of all, his great discovery, the Dirac equation, wasn't in fact done that way, although he afterwards maintained that it was. He invented the philosophy of the beautiful post partum. If you look at what he actually wrote at the time, it's clear he was heavily guided by the experiments. And afterwards, when he adopted this philosophy of the beautiful, he stopped making discoveries. It's sad that it didn't work. It didn't work for Einstein, either. Einstein had the same philosophy of the beautiful in later life, and it led him astray, too. The more brilliant you are, the more badly you get led astray. You couldn't find two brighter people than Einstein and Dirac. Both of them became sterile in the second part of their

---

12 In *Asimov Laughs Again* (Houghton Mifflin, 1992), Isaac Asimov claims to have made up the millihelen pun in the late 1940s. It is ironic that 1) Asimov was dying from kidney failure and heart problems while writing this humor book (his 470th and last book), and 2) this book is much funnier than his earlier humor book, *Isaac Asimov's Treasury of Humor* (Houghton Mifflin, 1971).

13 Paul Adrien Maurice Dirac (1902–1984) was a British physicist who won the 1933 Nobel Prize in Physics.

lives because they followed this philosophy. —Freeman Dyson, *The College Mathematics Journal,* vol. 25, no. 1 (January 1994), pages 20–21.

The ancients had a theory that the orbits of the planets were circles. Beautiful, but false. When theory does not match reality, it is the theory that must be rejected, not the facts of observation. When I'm in charge of a class, I try to predict how the students will react to certain lessons and how well they will do on my tests. Sometimes my predictions are terrible, and I have a tendency to say, "What is wrong with these students?" But that is like saying, "What is wrong with the planet Mars for not moving in a circular orbit?" Usually there is nothing wrong with the students. Following Dyson, our theories of learning and our expectations of student behavior must come from a close observation of the students we are currently teaching, not from dreamy recollections of our own student days, not from experience with other students in different times and places, and not from idealistic feelings about how students *should* act. If we do not base our expectations on close observation, then we can expect a violent clash with reality.

Indirect Proof

A man comes to a psychiatrist and claims to be dead. No matter what arguments the psychiatrist uses, the patient refuses to believe he is alive. Finally, the psychiatrist decides to try indirect proof.

"Do dead men bleed?" asked the psychiatrist.

"No, of course not. Everybody knows that dead men don't bleed," replied the patient.

"Well, then, if I prick your finger with a pin and if we squeeze out a drop of blood, will that prove you're not dead?"

"Yes," said the patient.

The psychiatrist found a needle, pricked the patient's finger, and squeezed out a drop of blood.

"Okay, you've convinced me that I was wrong," said the patient. "Dead men do bleed."[14]

---

[14] A version of this joke appeared in the journal *Etc.: A Review of General Semantics,* vol. XVI, no. 4 (Summer 1959), pages 458–459.

The previous joke is a nice way to introduce the topic of indirect proof. It can be followed by asking the class why famous explorers like Columbus and Magellan were so sure the earth was round. Asking a history question in math class will usually get their attention. One answer is in the following indirect proof.

Theorem. *The earth is not flat.*

Proof (indirect). Observe stars low on the northern (or southern) horizon. Then take a long northern (or southern) journey. These same stars will appear to rise above (or drop below) the horizon in the evening. When you return, the observers who stayed at the initial position will report that the stars did not appear to change their position. If the earth were flat this effect would not be observed. Since the effect was observed, the earth is not flat. Q.E.D.

Contrary to popular belief, many educated people in Europe in the Middle Ages, believed the earth to be round. I like to shock my students by telling them that the ancient Greek astronomers had seen a crude picture of the earth and therefore believed it to be round. Where, I ask the students, did they see this picture? See note [1].

Sometimes I will go off on a mathematical tangent and discuss things that only a few students appreciate. These tangents are important for the intellectual growth of the more talented students. But some students are bored or can't follow the discussions, and become frustrated. Sometimes a student will put up a hand and ask: "Is this going to be on the test?" Handling this question is difficult.[15]

The Power of Inverses

Whenever I teach the rule $a \log b = \log b^a$, I ask my students to use their calculators to tell me how many digits are in the decimal representation of 2^{1234}. That such an apparently hard question can be answered so easily with logarithms comes as a surprise. Solution: Set $10^x = 2^{1234}$ and take the $\log_{10}$ of both sides to get $x = 1234 \log_{10} 2 = 371.47\ldots$. Therefore, 2^{1234} contains 372 digits (not 371).

---

[15] A clever reply is, "The next test will cover everything you have ever learned in mathematics with special emphasis on what we learned since the last test." The problem with this reply is that it makes matters worse, not better.

Then I ask them: "Do you understand?" (Yes.) "Well if I give you another problem with different numbers will you be able to solve it?" (Yes, Mr. Stueben.) "Okay, then solve this one: How many digits are in the decimal representation of 2^{1234} in base 5?" This time they must solve $2^{1234} = 5^x$. They can take $\log_5$ of both sides and use the change-of-base formula, or directly take $\log_{10}$ of both sides. (Answer $=$ 532 digits.) With the right classes, these problems are a lot of fun.

Students know that $\exp(x) = e^x$ is the inverse of $\ln(x)$, therefore they are not impressed when I make a big deal of the statement $x = \exp(\ln(x))$. "So what, Mr. Stueben? Of course inverses cancel each other out." Then I point out $b^a = \exp(\ln(b^a)) = \exp(a \ln b)$, which gives us a simple way to calculate powers and roots in any computer language that contains a logarithm function and its inverse (at least up to round-off error).

The Irish mathematician Desmond MacHale wrote that certain techniques and tricks in mathematical proofs are themselves almost jokes. I think the previous trick of using inverses is almost a joke. Introducing inverses should leave an expression invariant—and it does—yet it also permits new manipulations. Two other *joke-techniques* are adding zero to a sequence (to make an even number of terms so that they may be paired), and adding 1 to and subtracting 1 from the same expression to change its form.

> In some colleges of music, part of the doctoral requirement is to compose an original full-length symphony. Because modern music sounds so weird, a good ploy is to take a well-known classical symphony, write it backwards and submit it as an original work. One student took the daring step of taking his professor's doctoral symphony and reversing it. He failed to receive his degree, the examiners remarking that he had reproduced Sibelius's Fourth Symphony with not a single note changed. —Desmond MacHale, *Comic Sections* (Dublin: Boole Press, 1993), page 32.

Whenever inverses are discussed, I like to tell that joke. A few students can guess the ending before the joke finishes. Others don't understand it until they review the nature of inverse operations or someone explains it to them. Sometimes humor helps makes a point or makes it more memorable or just adds a bit of fun to a topic. But no matter how potentially useful humor can be in teaching, it cannot be effective unless a teacher has previously established a good relationship with his or her class.

Practical Jokes You Can Play on Your Mathematics Professor[16]

1. Turn in all calculations worked in hexadecimal. Claim it's the notation of the decade.

2. At least once a week, ask how the lectures relate to Fibonacci numbers.

3. Ask questions that leave your professor speechless, such as: "What is the difference between a mapping and a lemma?", "But just what numbers are delta and epsilon?" See note [2].

4. Every time your professor asks "Are there any questions?" just say, "I don't understand." If your professor tells you that is not a question, then say again, "I don't understand."

5. The first time your professor uses $f(x)$, ask him why he is multiplying f times x.

6. The first time your professor uses the symbol $\therefore$ ask, "What is that?" When your professor says "therefore," reply: "Don't you mean *there are three*?"

7. Whenever the word "set" is mentioned always say "Yes, but what about the null set?"

8. Every time your professor explains something to you, say "I still don't get it."

9. Whenever your professor uses the word "proof," make a long groaning sound.

10. Keep a running total of the number of times your professor uses the word "hence." Turn in a weekly count to him. If your professor asks you why you are doing this, just say, "I thought you'd like to know."

11. At least once a week, ask in class why a member of the opposite sex isn't teaching the course.

12. Every time your professor makes an assignment, ask: "Why do we have to do this?"

13. Bring a graphing calculator to class. Anytime the professor writes down an approximation or draws a graph, work it out exactly on your calculator. As soon as you get an answer, announce that a mistake has been made.

14. INDIRECT PROOFS: Assume the opposite of what you are trying to prove and start reeling off every consequence that you can think of. Eventually an error will creep in. The error will lead to a contradiction, and you will be done. If the professor finds the error, then tell her she

---

16 Several of these were posted by different people on the sci.math newsgroup.

is being too picky and demand partial credit. If your professor tries to explain what you have really done, just keep saying, "I don't understand."

15. DIRECT PROOFS: Start with the given and begin drawing any conclusion that occurs to you. Stop when you are halfway down the page. Then write the final conclusion at the bottom. Begin working backwards and up the page. Be sure that every new statement contains enough information to support the statement below it. Eventually you will reach the middle of the page. You are done. Of course, there will be a HUGE gap between the two statements in the middle of the page. If the professor finds the gap, then claim all proofs have some gaps and demand partial credit. If your professor won't give it to you, then say, "I don't understand."

It is amazing how annoying the statement "I don't understand" can be to a teacher who has just spent considerable effort in explaining something. If the student could ask a specific question, then progress could be made. As a beginning teacher, I tried to insist that my students ask me questions so that I could determine where their misunderstanding lay. Incredibly, the students who told me they didn't understand had no questions either.

I think what happens is this: Some students find the material boring, or the lecture dull, or the material too difficult, or they are distracted by something in their thoughts. Then when the teacher asks a simple question, they do not say, "I haven't been listening to your lecture," but rather, "I don't understand." And of course they don't have any questions either. The first step toward helping such a student is to understand exactly what "I don't understand" really means. The second step is not to make matters worse with criticism or an impatient tone. But the best thing to do is to minimize the sleep-causing factors in the first place [3]. One day I realized that I could take a big step forward in being a better teacher just by avoiding the mistakes of a poor teacher [4]. The following is my personal list of teaching mistakes.

Teaching Mistakes

1. Giving too many passive lectures.
2. Omitting counterexamples and omitting examples of common errors.
3. Not offering suggestions toward efficient learning.
4. Not giving enough problems that use mixed strategies and require the recall of significant ideas.
5. Not checking homework daily.

6. Failing to show perspective, to show applications, and to explain connections with other parts of mathematics.

7. Not showing the internal motivation behind the mathematics and not answering the questions "Why would anybody want to learn this?" and "What can we do with this knowledge or skill?".[17]

8. Omitting discussions of the subject's history, its etymologies, and the personalities of its creators.

9. Omitting personal stories of my own experiences, my colleagues' experiences, and my former students' experiences in the study of mathematics.

10. Teaching only to the brightest[18] or ignoring the brightest.

11. Proving statements that are obvious to most students or giving proofs that few or none can follow.

12. Making the classroom experience equivalent to reading a textbook: definition, lemma, theorem, proof, corollary, definition, lemma, theorem, proof . . . .

13. Taking no interest in students' difficulties in mastering the subject.

Notes

1. During a lunar eclipse, earth's shadow appears on the moon's surface. This shadow is slightly rounded. Of course, that could mean the earth is a flat disk rounded at the edges. But in conjunction with other evidence, this fact helped convince educated intellectuals that the earth was most likely spherical.

2. According to Desmond MacHale, these particular questions were actually asked of university professors. Once in the third week of school in the middle of a pre-calculus lesson, a student asked me to define *domain*. A beginning teacher wouldn't know whether to laugh or cry at such a question. It took me 20 years to learn how to reply: "The domain of a function $f(x)$ is the set of all legal values for x." No criticism, no sarcasm; just answer the question and go on.

---

[17] Here is a comment from the Father of the C++ language: "I studied pure and applied mathematics [in Denmark]. This left me with an appreciation of the beauty of mathematics, but also with a bias towards mathematics as a practical tool for problem solving as opposed to an apparently purposeless monument to abstract truth and beauty. I have a lot of sympathy for the student Euclid reputedly had evicted for asking, 'But what is mathematics for?'." —Bjarne Stroustrup, *The Design and Evolution of C++* (Addison Wesley, 1994), page 23.

[18] "We teach toward the Einsteins and forget the shopkeepers. Mediocrity is unscholarly and unacademic, even though a small increase in ability across society may be more valuable than a few more people at the peak (who would probably have got there anyway.) —Edward de Bono, *Teaching Thinking* (Penguin, 1976), page 23.

3. "When a Student makes really silly blunders or is exasperatingly slow, the trouble is almost always the same; he has no desire at all to solve the problem . . . . Therefore, a teacher wishing seriously to help the student should, first of all, stir up his curiosity, give him some desire to solve the problem."
—George Pólya, *How To Solve It,* 2nd ed. (Doubleday, 1957), page 94.

4. For a long list of teaching errors from the student's viewpoint, see a letter from Olive Chapman in *The Mathematics Teacher,* vol. 89, no. 2 (February 1996), page 83.

3
Problem Solving

Mark Kac[1] (1914–1985) was a Polish mathematician and physicist who immigrated to the United States in 1938. He taught at Cornell University and did research at the Institute for Advanced Study in Princeton. The selection below was taken from his autobiography.

It was the summer of 1930 in Krzemieniec, Poland. I was sixteen. In September my last year in school would begin and it was time to think about choosing a career. My academic future, however, was not uppermost in my mind. I had been stricken by an acute attack of a disease which at irregular intervals afflicts all mathematicians and, for that matter, all scientists: I became obsessed by a problem. The symptoms are familiar and easily recognized, especially by the victims' wives, since they consist of a marked increase in antisocial behavior. Loss of sleep and appetite is also frequent. In my case the symptoms were especially pronounced; so much so, in fact, that my family was beginning to worry.

The problem that infected me with such virulence was actually of little significance and even lesser consequence. It concerned solving cubic equations and the answer had been known since Cardano published it in 1545. What I did not know was how to derive it.

The sages who had designed the mathematics curricula for secondary schools in Poland had stopped at solving quadratic equations. Questions by curious students about cubic and higher-order equa-

---

[1] Pronounced "cats."

tions were deflected with answers such as "This is too advanced for you" or "You will learn this when you study higher mathematics," thereby creating a forbidden-fruit aura about the subject. But I wasn't having any of this and was determined to find out how one goes about solving cubic equations.

Throughout my life I have had a number of bouts with the virus of obsession and a number of problems causing the infection turned out to be of some significance in mathematics and science, but at no time after the summer of 1930 have I worked as hard or as feverishly. I rose early and, hardly taking time out for meals, I spent the day filling reams of paper with formulas before I collapsed into bed late at night. Conversation with me was useless since I replied only in monosyllabic grunts. I stopped seeing friends; I even gave up dating. Devoid of a strategy, I struck out in random directions, often repeating futile attempts and wedging myself into blind alleys.

Then one morning—there they were! Cardano's formulas on the page in front of me. It took the rest of the day and more to pick out the thread of the argument from the mountain of paper. In the end the whole derivation could be condensed into three or four pages. —Mark Kac, *The Enigmas of Chance* (Harper and Row, 1985), pages 1–3.

This is the way science and mathematics are done. And this is why most people will never understand the mathematical experience. Who could possibly concentrate so long on one problem without becoming bored? Only someone with an abnormal psyche; only someone who has become obsessed. Normally this is not a good thing. But some problems are so difficult that only an obsessed person can muster enough thought and effort to solve them. This kind of obsession benefits society even though it plays havoc with an individual's personal life.

One of the reasons that people think problem solving can be taught is that they have personally experienced this kind of high-level learning. Revelations through private study, through discussions with friends, and through the long searching for answers are the joy of any intellectual. The discovery of unifying concepts and simplifying ideas is so wonderful that the discoverer wonders why these experiences are not more common in the schools. I have heard many intelligent people state that teachers must emphasize ideas over the memorization of facts. They say this because they recall that the best learning they experienced was the learning of ideas and the discussion of ideas. Why, they ask, is this not done more often? Are the teachers uninspired? Are they too dull to understand the ideas themselves? No. The answer is that deep learning requires a passionate interest on the part of the student. And although a student might be willing, he or she cannot always produce the necessary passion on demand. Consequently, building student interest in mathematics must be given a high priority.

Heuristics

Here is a poster I made up for my pre-calculus students:

Reminders

1. $\sqrt{x^2} = |x|$. An *even* power canceling with a radical introduces absolute value bars.
2. $|x| = 2 \Rightarrow x = \pm 2$. Removing absolute value bars around one side of an equation introduces the double sign ($\pm$) on the other side.
3. Never divide by x. Never divide both sides of an equation by a common variable expression, or a root may vanish. Instead move all terms to one side and factor out the expression.

I often tell students who have dropped roots in calculations to look at the poster and see how it applies to their work. Internet correspondent Eric Leong made up a charming mnemonic to train his students for math competitions. He asked his students to remember that their brains are *Air Force Base Computers,* where

> A is for Answer: Did you answer the right question?
> F is for Form: Is the answer in the right form?
> B is for Ball Park: Is the answer in the ball park?
> C is for Computations: Did you check your computations?

Simple advice written in humanistic terms is generally more memorable to the weaker student than precise textbook-like sentences. Here is another list:

High School Math Heuristics

1. Be sure you understand the problem. Note the domain and the constraints.
2. Draw a picture or diagram.
3. Look for a pattern.
4. Work backwards.
5. Examine special cases.
6. Calculate intermediate data, and then re-examine the problem.
7. Express the answer in terms of an equation, and then solve the equation.
8. Solve a related or similar problem that is easier.

9. Attempt to reduce the problem to a previously solved case.

10. Attempt to prove the problem impossible.

Lists like these are fun to construct, and they do have some value in focusing students' attention on certain key tactics. But these lists are of limited use. People who are good problem solvers use heuristics intuitively. And people who are weak problem solvers will only use heuristics if they have been connected by memory to certain problems. I think the teaching of problem solving is sometimes based on the idea that if we watch how successful students solve problems and teach the same techniques to weaker students, then the weaker students will be better problem solvers. No, they won't. In fact, the weak math-problem solvers might be made even weaker by such an experience. Teaching problem solving in this way is literally encouraging students not to think. A better way is to have students study their own errors.

In pre-calculus there is a theorem called the Remainder Theorem. It says that if polynomial $f(x)$ is divided by $(x - c)$, then the remainder will be $f(c)$. This theorem is handy for verifying factors of a polynomial. Once after teaching it, I tested with this question:

Find the remainder of $f(x) = x^{3700} + 2x^{3699} - 4x^{1786} + x^{1788} + x^7 + 794$ when divided by $(x + 2)$.

The students tried to substitute -2 for x and reduce the expression to a pure number. But their calculators could not handle a number as large as $(-2)^{3700}$, so some of them gave up. What I hoped they would see is that, after the substitution, the first two terms cancel each other, and the third and fourth terms cancel each other. This question requires students not to give up in the face of an apparent complexity. And if they do give up, they must account for their failure when they write up their test corrections. Some of my more serious students were noticeably disturbed by their inability to see the cancellation. A girl named Katherine came to me after the test and said, "I don't understand. I do the homework, I study for the tests, and I continue to make stupid mistakes on your tests." She was an extra-conscientious student who had received an A the previous quarter by doing a lot of extra credit. I told her that she was doing everything that she could, and that doing well on these tests was just a matter of time and maturing. I think the fact that she had been able to maintain an A kept her from giving up. Keeping student attitudes positive is vital to their success in learning. And when students complain or ask why the questions are difficult, or why their studying hasn't brought them better grades, then the time has come for empathy, gentle discussion of goals, encouragement, and compliments on their efforts.

In pre-calculus the students are told that complex factors of polynomials with real coefficients come in pairs (complex conjugates). If the coefficients are not real, then the complex factors will not come in pairs. Here is a test question I used:

> True or False: If $3+i$ is a root of the function $f(x) = ax^7 - 75bx^5 + 117cix^2 - dx + k$, where a, b, c, d and k are real coefficients, then $3 - i$ is also a root.

Many of the students did not inspect the function closely enough to spot the i hiding in the bushes. It's not enough to tell them: "Haste makes waste." "Read each problem carefully." "Check your work." "Look for traps." Because people don't learn important lessons from words; they learn from experience.

No doubt, over the years, most of my students will forget nearly every mathematical fact I teach them, and they will lose nearly every mathematical skill I help them build. So what will they carry away from the course? What benefit will they derive from our time together? Will they learn to prepare for their tests in life, to pay attention to deadlines, to check their work, to not give up easily, to communicate clearly, to show compassion to others, to think deeply, to try to be honest? Who knows? But these lessons are not part of any syllabus. They are learned through reflection on success and failure. They are learned through struggle in an environment that encourages intellectual and emotional growth.

Good Problems

> *Give a man a fish and you feed him for a day. Teach a man to fish and you feed him for a lifetime.*

I now want to show you the kind of problems that need to be presented in the upper-level classes. The first one was shown to me by "the guy down the hall," my fellow-teacher Steve Rose.

Problem 1. There is a single 21-foot high vertical palm tree growing in the middle of a large flat desert island. On a certain day, the sun will rise at 6:00 A.M. and set at 6:00 P.M. At noon on that day the sun will be directly overhead and the trunk of the palm tree will cast no shadow. A traveler lay down the night before this special day, and was awakened in the morning when the sun rays reached his eyes, which were 13 feet due west from the tree. What time did he awake? Round your answer to the nearest minute. The solution is in note [1].

This problem requires only a methodical approach with arithmetic and right-angle trigonometry. When I gave it to my freshmen (who worked it in groups), they took 15–20 minutes to solve it. With considerable effort on both the students' part and mine, about five out of six students solved it. My part consisted in walking around and saying, "Your reasoning sounds good so far," or "I don't think your diagram is correct," or "Sorry, you've made a mistake somewhere."

Problem 2. Tom Swift was testing a new experimental airplane. He departed from his airport and flew 1000 miles due north, 1000 miles due west, 1000 miles due south and 1000 miles due east. He landed exactly on the spot where he started. This was so much fun that Tom took another trip. This time he took off from the same spot as last time, but only flew 500 miles due north, 500 miles due west, 500 miles due south and 500 miles due east. But this time, Tom did not land on the same spot from which he departed. Where did Tom land and where on earth did he start from? Assume Tom was never closer than 200 miles to either pole at any time. The solution is in note [2].

Again, hints have to be given. A globe must be produced. A student in each group must notice that a 1000 mile east-west trip at one latitude is significantly different from another 1000 east-west (or west-east) trip at a different latitude. If no one in a group notices this fact, then it must be given to them so that they can get on with the rest of the problem.

Problem 3. Given two real numbers a and b, write an expression equal to the smaller of the two numbers. The solution is in note [3].

Shortly after the class has studied the PAI Theorem (if two parallel lines are cut by a transversal, then the alternate interior angles are congruent), I ask them to take out a clean sheet of paper and copy down the following information:

Problem 4. In one day (24 hours) the earth rotates 360° ____ degrees on its axis. In one year (365.22 days) the earth makes 365 ____ full turns on its axis plus part of another turn. [Please ignore the spaces after each number.]

Then I give them this little speech: "You are probably wondering why I had you write down facts that are known to every high school student in the world, facts that no educated person would ever deny. Our purpose today is

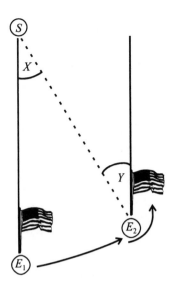

FIGURE 4

to prove these facts with geometry. And all we need to do is draw a simple picture."

Then comes the discussion: "The diagram [Figure 4] illustrates the same flag pole pointing directly at the sun at noon on two successive days. What is $\angle X$? How do you know? (Since there are 365.22 days in a year and there are 360° in a circle, $\angle X$ must be about 1°.) What is $\angle Y$? (Angle Y must also be about 1°) Why? (The PAI theorem). Hence, in one day the earth turns about 361° on its axis. *Students: Please change the first number in the box.*"

"Since there are 365 days in a year and the earth turns about 361° each day, we must conclude that in one year the earth makes ...(pause) ...how many full turns on its own axis? You tell me. (366 full turns on its own axis.) Students: Please change the second number in the box."

I follow this up by discussing how Eratosthenes determined the circumference of the earth around 240 B.C. [4]. Then they are asked to solve similar problems.

Problem 5. In the year 2200, two space explorers are assigned the task of measuring the circumference of a spherical asteroid. The first explorer flies up to the asteroid in her space scooter and places a vertical stick into the ground. She notes that the 10-inch shaft above the ground casts a 10-inch shadow. At the same time her co-worker (who is 100 miles away) does the same thing and notes that a) the shaft of the second stick casts a shadow in the same direction as the first stick, and b) the sun ray passing over the top edge of the second stick makes a 30° angle with the stick. Assume both sticks and the position on the asteroid with the sun directly overhead are all on the same great circle. What is the circumference of the asteroid? The solution is in note [5a].

The solution requires subtraction of the angles 45° and 30°. I assign problems that require subtraction, but then test on a problem that requires addition (where the sticks cast shadows in opposite directions on the same great circle). See note [5b]. Life meets our weaknesses, not our strengths. When the students ask me why I do this, I reply with the same philosophy I repeatedly state throughout the course: "It is a waste of your good time to solve problems only by memory and pattern recall. You must observe relationships. You must apply principles. You must think deeply. Someday you will be assembling something and the directions will ask you to connect part *D* with part *A*, but there will be no part *A*. What will you do then? Give up? No, you will think about how this could have happened; you will search for a different interpretation, or a mistake; and perhaps you will discover that part *H* is part *A*, because the top of the letter *A* didn't print out."

Most high school freshmen *do not know* that the equatorial circumference of the earth is about 25000 miles. Actually it is almost exactly 24900 miles. See note [6]. Nor do most of them know the number of days in each month. I require them to learn this knowledge for the next test. Some of them never learn it, or learn it for one test and then forget it for the next test. Anyway, after making a big deal about the number 25000 miles, I give them this extra-credit question on the next test:

Problem 6. Choose the closest answer to the following question: The diameter of the earth is approximately
A) 25 miles. B) 2500 miles. C) 8000 miles. D) 25000 miles.
E) 250000 miles.

Nearly everyone in the class chooses D) 25000 miles. (Wrong!) We learn best from our mistakes.

Kepler's second law (1609) states: *The earth does not move at a constant speed in its path around the sun. Rather, it moves so that the invisible line*

connecting the earth to the sun sweeps out equal areas in equal times.[2] This seems surprising, because the earth's orbit is not circular, and the sun is not at the center of earth's orbit. Every year I tell this to my geometry freshmen and then ask them the next question:

> **Problem 7.** What would happen if gravity turned off and the earth went flying off into space? (We assume the sun is fixed for the purposes of this problem.) Would Kepler's law still be true? Would it even make sense? The solution is in note [7].

I taught high school geometry from Moise[3]-Down's *Geometry* for more than a decade. The following problem appears in their third lesson on parallel lines in a plane. (The big theorem in that section is the PAI Theorem.) As is appropriate for a beginner's problem, the solution can be obtained easily by just looking at the diagram and perhaps making a few marks on it. But most students can't solve this problem. In fact, some geometry teachers can't get it in one sitting. I tell my students that if they can solve this problem, they will have surpassed some math teachers I have known. When the solution is given, there are plenty of groans at how obvious it is. I will never forget 14-year-old Rachel Lei turning to me (after she took my challenge and found a solution) saying, "That's a beautiful problem!"

> **Problem 8.** Given scalene $\triangle PMN$, $\overline{MX}$ bisects $\angle PMN$, $\overline{NX}$ bisects $\angle PNM$, and $\overline{QR}$, through X is parallel to $\overline{MN}$. If $PM = 10$ and $MN = 15$, and $PN = 17$, then what is the perimeter of $\triangle PQR$? [See Figure 5.] The solution is in note [8].

Students need a lot of help in getting through the next two problems. But this is a paradox, because in doing no more than examining and simplifying the form of each problem, the solutions drop out.

> **Problem 9.** For $x, y > 0$, prove: $\frac{x}{y} + \frac{y}{x} \geq 2$. Do not leave a radical sign in your final step. The solution is in note [9].

---

[2] Kepler discovered his three famous laws without using a telescope. In 1604 Kepler had shown that the intensity of light diminishes with the square of the distance from the light source. Later Newton would write: "If I have seen farther it was because I stood on the shoulders of Giants." Newton's Giants were Kepler, Galileo, and Descartes.

[3] Moise rhymes with the girl's name Louise.

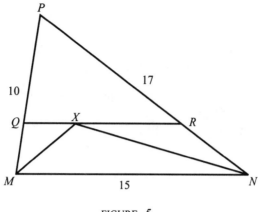

FIGURE 5

Problem 10. Prove $\log_{10} 5$ is irrational. The solution is in note [10].

Problem 11. You have 64 coins, all of which appear identical. However, one of the coins is heavier than the others. Using a two-pan balance, how many weighings do you need (in the worst case) to determine the false coin?

The students come up with an answer of 6 in a short time. But, I require the students to use their calculators to turn 64 into the answer. Most students recall my saying "binary division is log-base-2 work." They want to calculate $\log_2 64 = 6$ on their calculator, but there is no $\log_2$ key. Consequently, they apply the change of base formula: $\log_2 n = \log_{10} n \div \log_{10} 2$. After all this work I surprise them with the fact that 6 is not the answer! See note [11].

I learned a lot of mathematics from the following problem.

Problem 12. Solve algebraically for x: $(\log_{10} x^2)^2 = \log_{10}(x^4)$. The solution is in note [12].

The drill problems in most math textbooks are too easy and lack ingenuity. They are boring, one-step applications of the lesson presented on the same day. The problems are challenging only by being complicated, not by requiring insights and not by using a mixture of strategies. I think it is better to minimize feel-good mathematics and to offer more problems of substance.

This brings up the question of grading. It is better to grade too hard than too easy. The easy teacher will be more popular and will have less conflict with students, parents, and administrators. But the students will learn less from the easy teacher. This is because most students do not find school subjects intrinsically interesting enough to study on their own, and even students who do resonate with a subject will not study it to their full capacities due to the distractions of youth. Consequently, grades and deadlines are powerful learning motivators for high school students. In fact, they may be the two most important motivators in all of high school education.

> We now know that success in school leads to more positive self-esteem, not the other way around. Artificially inflating youngster's feelings of competence does little to promote genuine achievement and probably impedes it, since it erodes youngsters' sense of standards. Paradoxically, if we are genuinely concerned about improving the mental health of American youth, we ought to take steps to see that they are genuinely challenged and achieve more in school. — Laurence Steinberg, *Beyond the Classroom* (Simon & Schuster, 1996), page 96. [13]

Conclusion

I have talked much about maintaining good relationships with students. But this is not enough. The majority of high school students in the United States spend four or fewer hours *per week* on homework.[4] The teacher who does not collect homework daily is reinforcing poor learning habits and is making the job of teaching more difficult for other teachers. Homework of substantial intellectual challenge should be given every day. Since students will not do it without motivation, it needs to be counted as a considerable part of their grade. I count it as 40% of each quarter's grade. Nevertheless, about three weeks go by after the first assignment before I obtain 100% homework in even one class. Problems more difficult than those found in the textbook need to be assigned. I use foreign textbooks and older textbooks. There is a belief that a student who is forced to spend long hours on homework may not enjoy his youth. If the homework is approximately 30 minutes a night per class, my research indicates just the opposite is true. The engaged student seems to connect with life through his homework.

Major tests must be challenging, and they should contain some problems that the students have never seen before, but which are keyed to the principles

---

[4] Laurence Steinberg, *Beyond the Classroom* (Simon & Schuster, 1996), page 68.

taught in class. Test corrections must be required for all major tests, otherwise students will not learn from their mistakes. Requiring parent signatures on corrected tests is an effective motivator for future study. If a class does poorly on a topic, then problems covering that topic must continue to appear on homework and on tests until the concepts are mastered. The spiral method is the only method of teaching mathematics that makes sense in high school. Some timed memory quizzes should be given to force the memorization of necessary facts and formulas.

Discovery lessons, students writing to learn mathematics, the teaching of so-called general problem-solving concepts, field trips, math lab lessons, alternate assessments, collaborative partner tests, student presentations, and open-ended problems should all be used sparingly. I use some of them, but they have limited value. Pencil-and-paper analytic solutions are the heart of mathematics education.

There are two problems with this advice. First, the teacher must develop an enormous set of challenging drill problems. This takes years. Second, maintaining high standards and requiring students to be responsible may set the teacher in conflict with the students, the administration, the parents, and the department head. What will help is the maintenance of good relations with all involved. But in the end, one cannot stand alone against the world.

Notes

1. The answer to the palm tree problem is 9:53 A.M. Solution: 1) Draw the figure by constructing a right-angle triangle with vertical side equal to 21 and base equal to 13. 2) Recognize that the time of day before noon is directly associated with the angle of the sun above the horizon (= the acute base-angle of the triangle). 3) Calculate the acute base-angle of the triangle: $\arctan(21/13) = 58.24°$. 4) Using direct proportions, change the angle into hours and minutes: $58.24°/180° = x$ hours$/12$ hours $\Rightarrow x = 3.88$ hours. And 0.88 hours$/1.00$ hour $= y$ minutes$/60$ minutes $\Rightarrow y = 52.8$ minutes. So the answer is 6:00 A.M. $+ 3$ hours $+ 53$ minutes, or 9:53 A.M.

2. Tom started exactly 500 miles south of the equator. There is no other way to account for the fact that his first (4000 mile) trip departed from and arrived at the same location. He landed about 4 miles east of where he started. It took me about 45 minutes (using only arithmetic and plane trigonometry) to determine that the answer is 4 miles.

3. This problem is perfect for high school freshmen, because it requires only the notions of the distance formula for two points on the number line and the

midpoint formula. After they obtain the solution, it is a good idea to have them explain the solution in words. Answer: The coordinate of the smaller point is obtained by subtracting half of the distance between the two points from the coordinate of the midpoint.

$$\min(a,b) = \frac{a+b}{2} - \frac{|a-b|}{2}.$$

4. One day around 250 B.C., a librarian at Alexandria named Eratosthenes came across a curious fact about the North African city Syene (modern Aswan) which was about 500 miles south of Alexandria. He had been informed that on one-and-only-one day a year (the summer solstice), at noon, the sun comes directly overhead in Syene. Consequently, a vertical stick could cast no shadow at that moment. But at noon on the same day, a careful measurement of a vertical stick in Alexandria showed that the sun's rays made an angle of about one-fiftieth of a complete circle (7.2°) with the top of the stick. Assuming that the sun's rays are nearly parallel, and that the earth is nearly spherical, Eratosthenes applied a simple proportion to determine the circumference of the earth. (See Figure 6.)

5a. Solution: The circumference of the asteroid is 2400 miles. Consider a third stick (stick C) on the same great circle route as the other two sticks. Suppose this stick is directly under the sun and casts no shadow. Then stick A is 45° on the great circle away from C and stick B is 30° away from C. Since A and B cast shadows in the same direction, the two sticks are on the

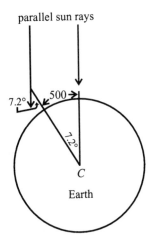

parallel sun rays

FIGURE 6
Eratosthenes's Diagram. $7.2°/360° = 500$ mi/x, where $x = $ circumference.
∴ circumference = 25000 mi.

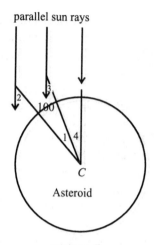

FIGURE 7

Space Explorer's Diagram. $\angle 2 = \angle 1 + \angle 3$, and
$\angle 3 = \angle 4 \Rightarrow \angle 1 = \angle 2 - \angle 3 \Rightarrow 15°/360° = 100$ mi/x, where x = circumference.
$\therefore$ circumference = 2400 mi.

same side of C. Since we assume the sun's rays are parallel, we may apply the
PAI theorem. Therefore, the sticks are 15° apart on the great circle. Since their
distance apart is given to be 100 miles, we have $15°/360° = 100$ mi/x mi.
5b.   If sticks A and B cast shadows in *opposite directions,* they would be
75° (= 45° + 30°) apart. In that case, the circumference of the asteroid would
be 480 miles.

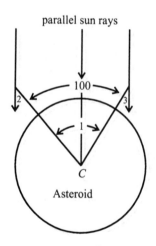

FIGURE 8

Math Test Diagram. $\angle 1 = \angle 2 + \angle 3 \Rightarrow 75°/360° = 100$ mi/x,
where x = circumference. $\therefore$ circumference = 480 mi.

6. In the *CRC Standard Mathematical Tables and Formulae* (1991), the polar radius of the earth is given as 6356.912 km, and the equatorial radius is given as 6378.388 km. Using CRC's conversion factor (0.6213711922 km = 1 mile) gives a polar circumference of 24818.59 miles and an equatorial circumference of about 24902.44 miles. There is a rumor that the 2.44 part is fictitious and was made up to circumvent the notion that 24900 is accurate to only the nearest hundred miles. Curiously, $\sqrt{5\pi}$ thousand mi = 24902.32 mi. The dimensions of the earth are calculated by assuming the earth is a perfectly smooth ellipsoid with its surface at mean sea level. Any question of the earth's size is complicated by the fact that the earth is slightly fluid (it has tides) and has land masses that are crinkly. Consequently, measures to the nearest meter are misleading. The average radius is nearly 3957 miles (*note:* all odd digits), and the average circumference is about 24860 miles (*note:* all even digits) or about 40009 km. Stating an average radius and an average circumference together may confuse students, because an average radius can not in general be used *to directly compute* an average circumference. Consider five squares with sides of $2''$, $3''$, $4''$, $5''$, and $6''$. The average side length is $4''$; the average area is $18''$, yet $4^2 \neq 18$.

7. If gravity were turned off, then Kepler's second law would still be true. Proof: In Figure 9, SX is $\perp$ to the line containing earth's new orbit. The area of each triangle is the same $\frac{1}{2}hb$. Q.E.D. This is a great question to ask on the day after the formula for the area of a triangle is introduced.

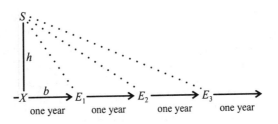

FIGURE 9

8. The perimeter is 27. Since $\triangle QMX$ and $\triangle RNX$ are both isosceles, we have $QX = QM$ and $XR = RN$. So the perimeter $= PQ + QX + XR + RP = (PQ + QM) + (PR + RN) = PM + PN = 10 + 17$.

9. For $x, y > 0$, prove $x/y + y/x \geq 2$. Add the fractions and remove the denominator obtaining: $x^2 + y^2 \geq 2xy$, or $(x - y)^2 \geq 0$. Most students want to quit here, but we can go one more step by taking the square root of both sides: $|x - y| \geq 0$. Since the steps are reversible, we are done. This solution leads to two important ideas: a) Radicals and fractional exponents of *real* numbers are always uniquely defined;[4] hence, the absolute value bars are necessary, and b) we may verify an identity over domain D by working on both sides simultaneously until we obtain a known identity *if and only if* all steps are reversible and do not restrict D. The verification is then obtained by reversing the steps. The following fallacies bring home these important ideas.

   a) $2 = \sqrt{2^2} = \sqrt{4} = \sqrt{(-2)^2} = -2$. Please insert the absolute value bars around the -2.

   b) Is $1 = -1$? Evidently it is: Square both sides and get $1 = -1$. Q.E.D. Alas, the steps are not reversible, and, of course, a false statement implies any statement.

10.   If $\log_{10} 5$ is rational, then $\log_{10} 5 = a/b$, where a and b are positive integers. Consequently, $10^{a/b} = 5$ or $10^a = 5^b$. But this is a contradiction, because all positive integral powers of 10 end in a zero, and all positive integral powers of 5 end in a 5. Q.E.D.

11.   The answer is $\lceil \log_3 64 \rceil = 4$ (note the ceiling function). Why? Because the solution can found using *trinary* division: Divide the coins into groups of 21, 21, and 22. Weigh the two groups of 21. If one pan dips, then the false coin must be in that pan. If the two pans balance, then the false coin must be in the group of 22, etc. My calculator gives int (x) $= \lfloor x \rfloor$ as the floor function; it has no key for the ceiling function. This leads to a nice extra credit problem: Write the ceiling function in terms of the floor function. Answer: $\lceil x \rceil = -\lfloor -x \rfloor$.

12.   Solve for x: $(\log_{10} x^2)^2 = \log_{10}(x^4)$. We must pay attention to the domains. Most algebra books state that for all positive real numbers x, n, and b, $(b \neq 1)$, $n \log_b x \Leftrightarrow \log_b x^n$. But in this problem x may be negative. Consequently, we derive a variation of the previous law: $\log_b x^{2n} \Leftrightarrow 2n \log_b |x|$, where n is a positive integer, and $x \neq 0$. So we have $(2 \log_{10} |x|)^2 = 4 \log_{10} |x|$. A simple substitution will reduce the number of symbols and thereby assist weaker students in seeing factors: Let $y = \log_{10} |x|$, and get $(2y)^2 = 4y$, or $y^2 = y$. Many students will now go wrong by dividing by y. *Never divide both sides of an equation by a common variable expression, or a root may vanish.* Instead, move all terms to one side and factor out the expression: $y^2 - y = 0$, or $y(y-1) = 0$. The solution is $x \in \{1, -1, 10, -10\}$.

13.   Laurence Steinberg's *Beyond the Classroom* is one of the best books I have read on explaining the factors that influence the academic performance of high school students.

---

   [4] This statement is true for real numbers, but not for complex numbers. The expression $z^{1/n}$ is *multivalued* for complex number z.

4
Computers and Calculators

In the 1992–1993 school year, my high school introduced the TI-81 graphing calculator into every mathematics classroom. As a joke, I told a few teachers that the TI-81 had a software bug in it, and it would not correctly graph some functions—e.g.,

$$y = \cos^2 x - 1 + \sin^2 x.$$

Get it? This is $y = 0$, and because the graph lies entirely on the x-axis, the user cannot see any points being displayed. Andy Samson, a teacher down the hall, caught the spirit of my humor and wrote me a note complaining that he found another bug: The calculator would not graph

$$y = \sqrt{\frac{x-1}{1-x}}.$$

I made copies of our equations and placed them in other teachers' mailboxes with a notice that the TI-81 calculator could not be trusted to graph equations correctly. Several days later Andy came to my room and told me that two teachers had taken him aside to explain that he was mistaken. There was nothing wrong with the calculator—his expression reduced to $y = \sqrt{-1}$, which has no graph in the real plane. When he told them that the two equations were a joke, they became angry. One of them complained: "From now on, anything I get from Mike Stueben is going directly in the trash can." I still haven't decided if my joke was a positive thing or not.

I sometimes give short quizzes that can be graded by just glancing at the answers. Each quiz is graded in front of the student who took it. If all the answers are correct, I place a HUGE 100% on the paper and give the student an enthusiastic compliment. If any questions are marked wrong, the student must return to his or her seat and correct them. This failure is accompanied by remarks, such as, "Oh, you really know how to do this one; you just made an arithmetic mistake" or "Your solution is probably correct for the picture you drew, but it's the wrong picture for this problem." Only the most confident students get teased: "Sorry, you have fallen into my trap for careless students."

Eventually most students get 100%, but a few students will not have finished by the end of the period. They must take their quizzes home and bring them back the next day. I tell them: "You are allowed to get help from your friends, but you are on your honor not to put down anything that you don't understand." These quizzes are especially welcomed by students at the end of the year when performance levels go down. But every year students complain that the questions are too hard. Let's take a look.

(1) Solve the following equation for x: $Bx^2 + Cx + A = 0$.

This is simply the quadratic equation with the letters permuted, and the solution, of course, is

$$x = \frac{-C \pm \sqrt{C^2 - 4AB}}{2B}.$$

But what a mess they make out of it. About half of the students are sent back to their seats.

(2) Find angle A in degrees if $\sin A = 12/13$ and $\cos A = -5/13$.

Most students just put SIN^{-1} (12/13) into their calculators and get 67.4°. (Wrong!) Quiz result: 33 correct answers and 22 incorrect answers. We learn best through our mistakes.

I once gave a quiz in which the first question required the students to place their calculators into radian measure and graph a function. The second question was to find the length of a vector. The third question was to calculate $\sin 20° \times \tan 30°$. The fourth question was about vectors again. Do you see the trap?

The students became distracted with question 2 and forgot to reset their calculators back into degrees. (I had repeatedly warned them about this.) Students argued that my answer key was in error. But I told them that I knew exactly what they had done wrong and sent them back to their seats to try again. Eventually a quiet scream was heard from a student who realized his or her error. Some of them handed me their papers with comments like "You're absolutely evil, Mr. Stueben." Great fun!

The students did poorly on these questions because the questions were designed to exploit common conceptual misunderstandings and lack of attention to important details. These questions were developed by a close examination of the errors made by students in previous years. Such drops of poison can be nuggets of gold in getting students to focus on their weaknesses.

Programming

What do you think of the following statement?

> High school programming is little more than carefully organized thinking.

Ask a class of high school programmers to write about this, and many of them will agree. But the above statement implies that just about anyone who can understand the syntax and execution of programming statements should be able to write high school programs simply by organizing his or her thoughts and by proceeding in a careful manner. I do not agree.

There are many intelligent and naturally well-organized students who find programming to be quite difficult. Programming requires a particular talent. Take the best high school programmers and ask their parents how well organized they keep their rooms. Ask their teachers how well they organize their time to complete assignments. Examine their English essays to see how well they have organized their thoughts. Ask the programmers themselves what systems they have adopted to keep from forgetting assignments and losing homework. Some of the best programmers seem to lack any other kind of organizational talent. The thinking behind the above statement is not only wrong, it's dangerous, because it leads some good programmers to think that other people are mentally defective. It leads some poor programmers to think that they are lacking intellectually.

> Trace the progress of the most ignorant of mortals, from his birth to the present hour, and you will be astonished at the knowledge he has acquired. If we divide all human science into two parts, the one consisting of that which is common to all men, and the other of what is peculiar to the learned, the latter will be insignificant and trifling in comparison with the other. —Jean Jacques Rousseau, *Émile, or A Treatise of Education, Book I* (1773).

In the process of teaching doubly subscripted variables in computer science, I have my students write the code to set all the elements in a 5×5 matrix (M) to zero. Most Pascal students write this:

```
FOR ROW := 1 TO 5 DO
    FOR COL := 1 TO 5 DO
        M[ROW, COL] := 0;
```

Next, they are required to write the code to fill the insides of the 5×5 matrix with 1s and make all the border elements equal to zero. The students immediately ask how many loops should be used. I tell them: "Some people use two loops, others use four loops, and some people include IF statements. Do it in the simplest way you can." Before an answer is displayed on the overhead screen, I insist on asking and answering the following three questions:

Question 1. How many operations are needed to place a hippopotamus into a refrigerator?

Answer 1. Three: Open the refrigerator; put the hippopotamus into the refrigerator, and close the door.

Question 2. How many operations are needed to place a giraffe into the refrigerator? (You may assume the refrigerator is taller than giraffe.)

Answer 2. Four: Open the refrigerator; take out the hippopotamus; put in the giraffe, and close the door.

Question 3. The giraffe and the hippopotamus are both thirsty and both one mile away from a river. The giraffe runs at a speed of 10 mph; the hippopotamus runs at half that speed plus 5 mph. Which one will get to the river first?

Answer 3. The hippopotamus will arrive first, because the giraffe is still trapped in the refrigerator.

With that behind us, I show them my code for putting 1s and 0s in the 5×5 matrix:

```
FOR ROW := 2 TO 4 DO
    FOR COL := 2 TO 4 DO
        M[ROW, COL] := 1;
```

There is a bit of a pause, and then somebody says, "Oh, I get it, the zeros are already on the border from the previous code." Yes. What has occurred before is relevant to the present.

Moral 1: If you think of a solution, look for a better one.
Moral 2: With matrices, remember the trick of overlaying values.

Serious Programming Jokes

Q: What is the most common exclamation heard in a computer science lab?
A: "Oh no! A special case."

Q: How many programmers does it take to change a light bulb?
A: Only one, but *nobody* is ever going to change that light bulb again.

A mathematician, an engineer, a physicist, and a programmer were all asked the same question by a psychiatrist: "What would you do if your car was broken?"

Mathematician: "I would buy a new car."

Physicist: "I would buy a new engine."

Engineer: "I would take the car to a service station."

Programmer: "I would drive the car around until it fixed itself."

A joke that has to be explained is a poor joke.—Old proverb. I consider the preceding jokes to be important because they each require the listener to recall some truth about programming to understand the joke. If the listener understands the joke, an important idea about style is reaffirmed with humor. This has more of an impact than just stating a rule about good style. If the listener doesn't get the joke, then he or she will ask for it to be explained. NEVER EXPLAIN JOKES WITH A POINT! If the joke remains a mystery, then the listener will reflect on it and may eventually solve it with some deep thinking. If it's explained, the listener will almost always feel the point of the joke is pedantic. Let me give you an example. There are certain stories called Zen parables and, it is claimed, these parables can be understood through experience, but cannot be explained. Here is one:

A Buddhist monk was being chased by a hungry lion. The monk came to the edge of a cliff and saw an old withered branch sticking out. So he climbed out on the branch just as the lion arrived. But then he heard a sound below him. It was a bear waiting for the branch to break so that he could eat the monk. Suddenly two squirrels appeared on the branch and started gnawing at its base. The monk looked over at the cliff and saw some wild strawberries growing there. He pulled off one, popped it into his mouth, and said, "Wonderful! I've never tasted berries this good."[1]

I like to tell this story to my students when they are complaining about how difficult my last test was. "What does it mean?" they ask. Zen parables can be explained, but it is always a mistake. In the past I did try to explain to my students what it meant to me: That even in the midst of difficulties we need not be miserable, and we need not forego moments of pleasure. But high school students are not experienced enough to know this as truth, and they think that it must be a poor story if it can't be understood without an explanation. It is better to let them reflect on the mystery. A good way to finish these jokes-with-a-point is by saying "class dismissed."

Assignment: Obtain an elephant from Africa.

Physicist: Starting on the west coast, he searches north-to-south and south-to-north slowly moving east. He inspects all gray animals keeping the first one that weighs the same as a known adult elephant ±500 pounds.

Mathematician: Starting in the center, he moves in an elliptical spiral (with major axis oriented north-to-south) removing all non-elephants and keeping whatever is left.

Computer scientist: First he notes that there are two kinds of elephants (African and Indian) and requests more detailed specifications as to which elephant is desired to be captured. Then he searches east-to-west and west-to-east starting from the southern tip and moving north. He stops only when encountering and capturing an animal whose description matches the American Zoological Society's classification of the type of elephant he is seeking.

Experienced computer scientist: Same as inexperienced computer scientist, except that he places a known elephant in Cairo to guarantee that the algorithm terminates.

Assembler programmer: Same as the experienced computer scientist, but he does it on his hands and knees.

I like to tell this joke when I teach the insertion sort. My version of the insertion sort first locates the smallest element and then swaps it into first place, thereby placing an elephant in Cairo. Why? So that each pass of the insertion algorithm will terminate before referencing the nonexistent zeroth position.

---

[1] Actually, this story has more to do with Tao than Zen.

If This Be Recursion . . .

A college professor once offered the following creative final exam: Write a suitable final exam for this course and supply a key. The first paper handed in read "Final Exam: Write suitable final exam for this course and supply a key. Key: Any reasonable variation of the previous sentence = 100%."

The world is divided into two classes: The people who don't divide the world into two classes and the people who divide the world into two classes: The people who don't divide the world into two classes and the people who divide the world into two classes: The people who don't divide the world into two classes and the people who divide . . .

LOOF LIRPA WINS CONTEST![2]

First prize in the Tribune's contest for the most interesting, yet possible, headline-with-following-sentence goes to Mr. Loof Lirpa for his entry: "Loof Lirpa Wins Contest! First prize in the Tribune's contest for the most interesting, yet possible, headline-with-following-sentence goes to Mr. Loof Lirpa for his entry: "Loof Lirpa Wins Contest! First prize in the Tribune's contest for the most interesting, yet possible, headline-with-following-sentence goes to Mr. Loof Lirpa for his entry: . . .

Professor Littlewood wrote a paper for a French mathematical journal. Professor Riesz translated it for him into French. At the end were three footnotes all in French: The first sentence read, "I am greatly indebted to Professor Riesz for translating the present paper." The second one read, "I am indebted to Professor Riesz for translating the previous footnote." The third one read, "I am indebted to Professor Riesz for translating the previous footnote." Littlewood then asks if he may legitimately stop here. What do you think?[3]

---

2 This is my variation of Littlewood's joke in his wonderful *Littlewood's Miscellany* (ed. Béla Bollobás, Cambridge, 1986), page 58. I think this is the funniest math book ever written on a serious level. Read this book! "Loof Lirpa" is "April Fool" spelled backwards.

3 As Littlewood later said: "However little French I know, I am capable of copying a French sentence."

On Method

I like to introduce this kind of humor when I am teaching recursion. Other teachers will wisely reject what does not fit well with their philosophy and style, and what does not work well with their students. The fanciful notion that what makes teacher A successful will consequently make teacher B successful is unfounded and misguided. Note A.P. Southwick's observation made in 1887:

> [T]he cannons of pedagogy are not fixed and immutable like those of mathematics, and there are opposing views on all subjects here discussed ... The teacher should know that there is pre-eminently no "The Method." Method is the outgrowth of philosophy, and must adjust itself to the laws of the mind and to the exigencies of science. The power of every true teacher is in himself, his personality, his character, his spirit and his attainments. —A.P. Southwick, *A Quiz Book on the Theory and Practice of Teaching* (Syracuse, NY: C.W. Bardeen, 1887), preface.

In our school there is a well-respected biology/earth-science teacher named Dennis McFaden. One time after lunch I dropped by to tell him a joke and noticed a lot of student activity. I asked what his students were up to. He then took me on a tour and explained that the students had been divided into groups and had accumulated earth and rock samples from around the county. They were performing chemical experiments on those samples and studying them under a microscope. They were drawing conclusions about the age of the local environment and under what conditions it was formed. I was impressed by how much he was teaching them and how enthusiastically they worked.

About a month later McFaden and I were talking about teaching and, on a whim, I asked him to tell me a truism about teaching. He replied, "Okay, I'll give you a truism. If you can make your students love you, then you can take them much further." Considering the kind of success McFaden had achieved with his students, I felt that I was being told one of the great truths of teaching by a practicing master. But, on reflection, I realized that it was a truth I could never apply. I simply cannot bond with my students the way McFaden does. And if I tried, the students would ultimately discover I was patronizing them. There may be some "best ways" of teaching, but these ways are not open to everyone. In that sense there is no "one method" for all teachers. It is also the case—I'm sure—that McFaden will always be a better teacher than I am. But it's one of those things that doesn't matter in the long run. What matters is that I reach a certain level of effectiveness.

The Bubble Sort

Here is my code for the infamous bubble sort:

```
FOR N := 1 TO MAX DO
    FOR M := 1 TO MAX-N DO
        IF X[M] >X[M+1] THEN
            SWAP(X[M], X[M+1]);
```

Suppose $n = 101$ and the elements are given in reverse order. Then this algorithm will make $100 + 99 + 98 + \cdots + 3 + 2 + 1$ swaps. Therefore, the number of swaps done by this sort (in the worst case with n elements) is proportional to the sum of an arithmetical sequence. Before explaining to a class how to derive the sum of these numbers, I tell them that the German super-mathematician Carl Gauss (1777–1855) was presented with the same problem when he was in elementary school. His teacher assigned his class the task of summing the integers from 1 to 100. The teacher was probably expecting some minutes of quiet, but Gauss produced the answer in seconds. Exactly which of several possible methods Gauss used I'm not sure, but I suspect that he divided the set of 100 numbers into 50 pairs, each of which summed to 101: $1 + 100, 2 + 99, 3 + 98, \ldots, 50 + 51$. Hence, the answer is $5050 (= 101 \times 50)$. Fine, but suppose there were an odd number of addends. No problem. Gauss could have appended zero, and then there would have been an even number of addends.

Using Gauss's trick, the students can now derive the general formula for the sum of the first n integers: $(n/2) * (n + 1)$. Thus we have proved that the time it takes to do a bubble sort (in the worse case) is asymptotically proportional to the square of the number of elements ($t = kn^2$)—a nice proof that the bubble sort is an n-squared sort in the worst case. But there is another way to obtain the required expression:

$$
\begin{array}{rcccccccccc}
T = & n & + (n-1) & + (n-2) & + \cdots + & 3 & + & 2 & + & 1 \\
T = & 1 & + 2 & + 3 & + \cdots + & (n-2) & + & (n-1) & + & n \\
\hline
2T = & (n+1) & + (n+1) & + (n+1) & + \cdots + & (n+1) & + & (n+1) & + & (n+1)
\end{array}
$$

Therefore, $2T = n(n + 1)$, or $T = (n/2) * (n + 1)$. Q.E.D.

I write out the first line and ask the students to copy it into their notes or scratch paper. Then I give them these directions: "Solve for T by reversing the right-hand side and adding it to itself." They are confused, so the directions have to be repeated and explained. Then students will have to be told to "simplify more" and "simplify again." Eventually some of the students get it and immediately recognize that they have derived the previous formula in a

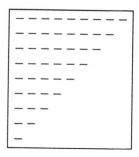

FIGURE 10

new way. A nice surprise. Finally, I ask them if the above picture is a proof that the bubble sort is an n-squared sort. I believe it is.

Now back to Gauss and the rest of the story.

School master: Class, I want each of you to spend the next quarter-hour summing the numbers from 1 to 100. This is difficult and I'm not sure of what your chances are for success.

Gauss: Fifty-fifty?

School master: Incredible! You calculated the correct sum immediately! You're a genius!

Non-serious Computer Jokes

Q: What is the one thing you should never say to your computer science instructor?

A: "Hey, watch this run!" It is inevitably followed by, "Oh no! What happened?"

The "slow sort" (or "permutation sort") is rediscovered every year by some of my students. This sort consists of exchanging two randomly selected items in an array and then checking to see if the array is sorted. If not, then repeat the process until the array is sorted. It is just the sort you want to use when your boss sends you to Paris and says, "I don't care how long it takes you to do it, just get the job done."

Q: Why does it always take exactly π times longer to write a computer program than is expected?

A: Because in the debugging process there is always a certain amount of going around in circles.

My former student Karl Kruger once used the variable names SNUE and UPDOK for one of my programming assignments. Looking over his program I asked: "What's SNUE?" ("Oh nothing, Mr. Stueben, What's new with you?").

More Dirty Tricks

In computer science the students learn binary, octal, and hexidecimal arithmetic. They learn that special symbols must be introduced to denote digits beyond 9 in base 16. When the test arrives they find this question:

Test question:
$$\frac{FAO_{16}}{10_2} = ?_{2000}.$$

When students see this they are shocked and ask if a mistake has been made. They have never worked in base 2000 before. What they need to do is to get over their shock and begin working, because the problem is straightforward. It just looks strange. And this is why the problem is important: It teaches a lesson in motivation that applies to problem solving in general. See note [1].

The second lab I usually give in computer science is to translate the quadratic formula into a line of Pascal code. The solution for one root is

```
X1 := (-B + SQRT(B*B-4*A*C)) / (2*A)
```

This lab has some great traps for beginning freshmen: They will write 4AC for 4*A*C, they divide by 2*A instead of (2*A), and they forget the outer parentheses around the numerator. When they get stuck and ask for help, I tell them that the whole point of the lab is for them to make the translation without help from me or their classmates. I am hoping they will try the following: 1) Double check their work and pay close attention to details. 2) Rewrite the expression from scratch. 3) Rewrite the expression a different way. 4) Work backwards: Translate their faulty Pascal expression back into algebra to find out which parts are different. 5) Work through it by baby steps: First write B*B and evaluate it, then write B*B-4*A*C, and evaluate it, then SQRT(B*B-4*A*C), etc. This is one of the most effective techniques for locating a programming error.

Unfortunately, the best laid plans of teachers often go awry. The weaker students become frustrated: They are afraid to fall behind, they are afraid to get a low grade, and they want to be working on the next lab like their friends.

So they peek at other student's solutions or ask friends to tell them what is wrong. I discuss all of this with them, but they are immature and school pressures are hard to ignore. My policy is to forgive all who transgress, and then move on.

Here are two more of my dirty tricks:

```
GIVE THE OUTPUT:
A := 1;
B := 2;
IF A >B THEN IF B = 1 THEN WRITE ('NICE')
          ELSE WRITE ('BAD');
```

Well, BAD is not the answer. And NICE is not the answer either. So what is the answer? The correct answer is "no output." This is the notorious case of the dangling ELSE. In Pascal the ELSE goes with the most recently executed IF. That fact combined with misleading indenting is why they get it wrong. Telling them to be on the lookout for this trap is not enough.

Here is the second trick. Give the output for the following Pascal code:

```
{--Initialize- - - - - - - - - - - - - - - - - - - - - - - - - - - - - - - - - -}
    FOR N := 1 TO 9 DO X[N] := N;
    X[10] := 5;
    LARGEST_X_VALUE     := 0;
    POSITION_OF_LARGEST := 0;
{--Search- - - - - - - - - - - - - - - - - - - - - - - - - - - - - - - - - - - -}
    FOR N := 1 TO 10 DO BEGIN
        IF X[N] >LARGEST_X_VALUE THEN
            LARGEST_X_VALUE     := X[N];
        IF X[N] >LARGEST_X_VALUE THEN
            POSITION_OF_LARGEST :=   N;
        END;
{--Print largest- - - - - - - - - - - - - - - - - - - - - - - - - - - - - - - -}
    WRITELN('THE LARGEST VALUE IS',    LARGEST_X_VALUE    );
    WRITELN('IT IS FOUND IN POSITION', POSITION_OF_LARGEST);
```

This problem was placed on one side of a worksheet. The other side contained a set of drill problems due the next day. Each student was required to get this problem correct before beginning the other side. I walked around the room waiting for answers. When the first student solved it, I said, "Scott got it. Who'll be next?" In a few moments: "Alice got it, too. Scott and Alice are the two class geniuses today." The problem is tough. I had to start giving hints. I passed one girl who wasn't trying—she was drawing a picture of a dog. "Pretty close, Mindy, you almost have it." We both chuckled. After a couple of minutes, I asked someone to give the correct answer and explain the trap. See note [2]. This little exercise encapsulates the essence of teaching as I see it, which is giving most students a positive experience—e.g., interesting and challenging.

The intellectual part is easy. Just by being in the teaching business, teachers run across good problems. But causing students to have a positive experience in class is where the teaching part comes in. And the key that allows a teacher to bring students positive experiences repeatedly is the rapport that is maintained with the students.

In my first year of high school teaching, my fellow teacher Gene Sabo told me how to handle a brainy, smart-alecky kid in my pre-calculus class. "The moment Eddie criticizes you, just come down on him hard." I did and it worked. Thanks, Gene. I even worked on rebuilding our relationship, but eventually he did more goofy things than a beginning teacher could handle, and Eddie and I parted ways not liking each other. What I needed to tell Eddie—repeatedly—was how to handle me in class (actually how to behave in the presence of any authority). The problem was that I kept getting angry with him and, instead of helping him with our relationship, I took the quick fix: I put him down in class and then went my own way.

No one can give you explicit directions for building and maintaining good relations, but I can tell you what derailed me for years. I became distracted by students not doing their homework, making rude remarks in class, and ignoring me when I made efforts to help them. I would become angry and resentful. If I thought about these students at all, it would be in critical ways. What wasted opportunities!

Scan the following list of activities.

Teaching Activities

1. **Prepare the next day's lesson:** Skim the section in the book, choose some good examples for the topic, prepare a warm-up problem that will review or lead into today's lesson, photocopy a worksheet and three-hole punch the copies, choose problems to assign for homework, and sort papers to hand back.

2. **Prepare a quiz:** Make out a key and photocopy the quiz.

3. **Update gradebook:** Grade recent quiz, record answers, and check in the daily homework.

4. **Attend meetings:** Faculty, departmental, subject-team, and committee meetings.

5. **Teach Class:** Take roll, give warm-up, pass back papers, go over homework answers, give the daily lesson, and give help to individuals.

6. **Tutor:** Tutor during activity periods, during class, during study hall, after school, before school, and sometimes during lunch.
7. **Straighten room:** Straighten files, report broken chair, clean up desk, and get supplies.

If you bother to read a book on teaching high school mathematics, you will find that most of the above activities are discussed along with other topics such as classroom management and suggestions on how to teach certain topics. All of this has some importance, but most of it can be figured out on the job, or by recalling how our own teachers taught, or by watching other teachers. All of these activities are on the periphery of teaching. Teaching is not even the organized presentation of material. That's the job of a textbook. *Giving experiences and keeping students focused is the job of a teacher.*

Mathematics (beyond arithmetic and descriptive geometry) is the most abstract of all high school subjects. One needs a talent for it to appreciate it. And if a student is not occasionally impressed with its ingenuity, its applications, its solutions by analogy, and by its shifts in perspective, then the requisite drill can become meaningless and boring. The average student cannot appreciate mathematical subtleties without the help of a knowledgeable teacher. The average student will not remain virtuous in study without the guidance of a watchful and involved teacher. Mathematics, more than any other subject, must be taught with the student in mind.

Notes

1. $\dfrac{FAO_{16}}{10_2} = 102_{2000}$. This problem was invented in 1987 by my student Jeremy Goldberg.

2. The answer is 9 and 0. The second IF never gets executed.

5
Mathematics Education

The following criticisms are mostly cynical, uncomplimentary, and, I believe, truthful. In a different time, I might have been fired and blacklisted for distributing this material.

H.L. Mencken on Education[1]

Next to the clerk in holy orders, the fellow with the foulest job in the world is the schoolmaster. The schoolmaster's [business] is to spread the enlightenment, to make the great masses of the plain people think—and thinking is precisely the thing that the great masses of the plain people are congenitally and eternally incapable of.

The art of pedagogics becomes a sort of puerile magic, a thing of preposterous secrets, a grotesque compound of false premises and illogical conclusions. The aim seems to be to reduce the whole teaching process to a sort of automatic reaction, to discover some master formula that will not only take the place of competence and resourcefulness in the teacher but that will also create an artificial receptivity in the child. Teaching becomes a thing in itself, separable from and superior to the thing taught.

---

[1] Selected parts from H.L. Mencken, *A Mencken Chrestomathy* (Knopf, 1953), pages 301–306. First printed in the *New York Evening Mail,* Jan. 23, 1918.

That ability to impart knowledge, it seems to me, has very little to do with technical method. It consists, first, of a natural talent for dealing with children, for getting into their minds, for putting things in a way that they can comprehend. And it consists, secondly, of a deep belief in the interest and importance of the thing taught, a concern about it amounting to a kind of passion.

This passion, so unordered and yet so potent, explains the capacity for teaching that one frequently observes in scientific men of high attainments in their specialties—for example, Huxley, Ostwald, Karl Ludwig, Jowett, William G. Sumner, Halsted and Osler—men who knew nothing whatever about the so-called science of pedagogy, and would have derided its alleged principles if they had heard them stated. It explains, too, the failure of the general run of high-school and college teachers—men who are competent, by the professional standards of pedagogy, but who nevertheless contrive only to make intolerable bores of the things they presume to teach. No intelligent student ever learns much from the average drover of undergraduates.

The truth is that the average schoolmaster, on all the lower levels, is and always must be essentially and next door to an idiot, for how can one imagine an intelligent man engaging in so puerile an avocation?

Go back, now, to the old days. Penmanship was then taught, not mechanically and ineffectively, by unsound and shifting formulae, but by passionate penmen with curly patent-leather hair and far-away eyes. Asses all! Preposterous popinjays and numskulls! Pathetic imbeciles! But they loved penmanship, they believed in the glory and beauty of penmanship, they were fanatics, devotees, almost martyrs of penmanship—and so they got some touch of that passion into their pupils. Not enough, perhaps, to make more flourishers and bird-blazoners, but enough to make sound penmen.

Such idiots, despite the rise of "scientific pedagogy", have not died out in the world. I believe that our schools are full of them. There are fanatics who love and venerate spelling as a tom-cat loves and venerates catnip. There are grammatomaniacs; schoolmarms who would rather parse than eat; specialists in an objective case that doesn't exist in English. There are zealots for long division, experts in the multiplication table, lunatic worshipers of the binomial theorem. But the system has them in its grip. It combats their natural enthusiasm diligently and mercilessly. It tries to convert them into mere technicians, clumsy machines. It orders them to teach, not by the process of emotional osmosis which worked in the days gone by, but by formulae that are as baffling to the pupil as they are paralyzing to the teacher.

We cannot hope to fill the schools with persons of high intelligence, for persons of high intelligence simply refuse to spend their lives teaching such banal things as spelling and arithmetic. Such men would not only be wasted at the job; they would also be incompetent. The business of dealing with children, in fact, demands a certain jejunity of mind. The best teacher, until one comes to adult pupils, is not the one who knows most, but the one who is most capable of reducing knowledge to that simple compound of the obvious and the wonderful which slips into the infantile comprehension. The best teacher of children, in brief, is one who is essentially childlike. — H.L. Mencken, 1918.

Selected Comments on Mathematics, Mathematicians, and Education

Let me, for a moment, become a bit more personal. Having associated from early years with two particular classes of scholars— botanists (or more widely, nature lovers) and mathematicians—I came to notice, and through the years have confirmed, a striking general difference between these two classes. The botanists are usually the more pleasant sort of people to be with; they radiate gentle modesty, are open-minded, enjoy one another's company, are kind in their professional comments about one another, and are found interesting by their nonbotanical friends. The mathematicians, on the other hand, are too often unpleasant to be with; they frequently exude self-importance, are professionally opinionated, tend to bicker and quarrel among themselves and to say unkind things of one another, take an almost gleeful pleasure in unearthing an error in another's work, and are often quite boring to their nonmathematical acquaintances. —Howard Eves (1911–), *In Mathematical Circles* (Prindle, Weber & Schmidt, 1969), page 85.

[I]t is easy to see that people in general divide into those with a bent toward numbers and science and those who take naturally to literature, history, and the arts. ... What is seldom noticed is that the trait in students who favor math is also the cause of much poor teaching. For the mind that feels at home with math is likely to lack the qualities that make a good teacher. ... [T]he easy handling of quantity is either an acquired habit or a taste that often implies a retreat from human contacts and the concerns of life. The same type of personality—silent, abstracted, seemingly dour or unfriendly—is often found among the scientifically inclined. That is how the math or science teacher may fail, although well prepared and of course well intentioned. He or she, moving easily among abstractions and their symbols, has no idea how the "other mind" works—does not perceive how strange the square root of minus one seems to the concrete minded, does not grasp where the difficulties lie, does not know what to take or not to take for granted in matters long obvious to the born quantifier. These are failings in pedagogy, not in knowledge. It is a case of two minds out of tune with each other. — Jacques Barzun, *Begin Here, the Forgotten Conditions of Teaching and Learning* (Chicago University Press, 1991), pages 78–79.

Un mathématicien de plus, un homme de moins. (One mathematician more, one man less.) —Félix Dupanloup (1802–1878), French educator, prelate, and statesman.

Abstract work, if one wishes to do it well, must be allowed to destroy one's humanity; one raises a monument which is at the same time a tomb, in which voluntarily, one slowly inters oneself. — Bertrand Russell (1902), in *The Autobiography of Bertrand Russell 1872–1914* (Grosset & Dunlap, Bantam Books, 1969), page 217.

What triumphed at the time [1960] is an idea that still holds sway in mathematics departments today [1987], namely, the simplistic view of mathematics as a linear progression of problems solved and theorems proved, in which any other function that may contribute to the well-being of the field (most significantly, that of exposition) is to be valued roughly on a par with that of a janitor. It is as if in the filming of a movie all credits were to be granted to the scriptwriter, at the expense of other contributors (actors directors, costume designers, musicians, etc.) whose roles are equally essential for the movie's success. —Gian-Carlo Rota, *American Mathematical Monthly*, vol. 94, no. 7 (August-September, 1987), page 701.

At the beginning of this century a self-destructive democratic principle was advanced in mathematics (especially by Hilbert), according
to which all axiom systems have equal right to be analyzed, and
the value of a mathematical achievement is determined, not by its
significance and usefulness as in other sciences, but by its difficulty
alone, as in mountaineering. This principle quickly led mathematicians to break from physics and to separate from all other sciences.
In the eyes of all normal people, they were transformed into a sin-
ister priestly caste ... Bizarre questions like Fermat's problem or
problems on sums of prime numbers were elevated to supposedly
central problems of mathematics. —V.I. Arnold, *The Mathematical
Intelligencer,* vol. 17, no. 3 (1995).

It is surprising how much those who represent the mathematics
department on university-wide committees and legislative bodies
are not representative of the department—the research people won't
take those assignments. —Harley Flanders, *American Mathematical
Monthly* (March 1971), page 291.

There was another bookish lad in the town, John Collins by name,
with whom I was intimately acquainted. We sometimes disputed,
and very fond we were of argument, and very desirous of confuting
one another, which disputatious turn, by the way, is apt to become
a very bad habit, making people often extremely disagreeable in
company by the contradiction that is necessary to bring it into practice; and thence, besides souring and spoiling the conversation, is
productive of disgusts and, perhaps, enmities where you may have
occasion for friendship. I had caught it by reading my father's books
of dispute about religion. Persons of good sense, I have since observed, seldom fall into it, except lawyers, university men, and men
of all sorts that have been bred at Edinborough.

Thomas Godfrey, a self-taught mathematician, great in his way, and
afterward inventor of what is now called Hadley's quadrant. But
he knew little out of his way, and was not a pleasing companion;
as, like most great mathematicians I have met with, he expected
universal precision in everything said, or was for ever denying or
distinguishing upon trifles, to the disturbance of all conversation. He

soon left us. —Benjamin Franklin, *Autobiography* (Bantam, 1982), pages 14, 54.

If you look up cellular automata on one of these computer searching things you'll find that there had been about a hundred papers written about them by 1981 or something, and so I went and looked up a whole bunch of these things, but they were boring. They were so boring! They were an illustration of a sad fact about science, which is that if someone comes up with an original idea, then there will be fifty papers following up on the most boring possible application of the idea, trying to improve on little pieces of details that are completely irrelevant. —Stephen Wolfram, in Ed Regis, *Who Got Einstein's Office* (Addison Wesley, 1987), pages 237–238.

In an article in the *American Mathematical Monthly* (September 1969), Professor I.N. Herstein of the University of Chicago pointed out that 75 percent of the students trained to do mathematics research never do so after acquiring the Ph.D.; they become teachers at two- and four-year colleges that do not demand research. Other studies confirm this fact. Moreover, surveys made of the research done by Ph.D.'s report that fewer than 20 percent published even one paper a year, and this says nothing about the quality of those papers. —Morris Kline, *Why the Professor Can't Teach* (St. Martin's Press, 1977), pages 245–246.

What Makes A Good Math Teacher?

What makes a good math teacher? The Mathematical Association of America posed this question to undergraduates at different universities and colleges. The responses of nine mathematics majors were published in the November 1994 issue of *Math Horizons* (pages 20–23). I have taken a sentence or two from each response. Do you notice a pattern?

Math teachers should be examples to students, and that includes relating to people of different ethnicity, gender, and background. —Jill Clasen, Senior at Clarke College.

First, and most importantly, good teachers have the ability to relate to their students. —Mark D. Drake, Senior at University of Nebraska-Lincoln.

The first quality a math professor must have is a degree of patience. A good teacher also must be able to understand students as individuals. —Steve Goodman, Senior at University of Dayton.

The love and passion for mathematics is what makes a great math teacher. —Dyronne Regis Luarca, Sophomore at University of San Diego.

In my experience, what makes a good math teacher has less to do with what a person teaches you than with how they go about teaching you. I believe that virtually anyone can teach you the basics of reading, 'riting, and 'rithmetic, but it takes an extraordinary person to make a student feel special, intelligent and cared about. —Rachel Mermel, Junior at Lake Forest College.

They should be approachable. Teachers should care genuinely for students, not just in an academic sense, but as people. —Jennifer McMillan, Sophomore at Hendrix College.

Pursuing two majors, mathematics and communications, has given me an opportunity to encounter numerous teachers. Communication teachers readily accept creativity and new ways to look at things. Mathematics teachers don't always respond so positively. —Steffanie A. Ries, Junior at University of San Diego.

I feel that the best math teachers have the ability to relate to students. —Michele Roll, Junior at St. Bonaventure University.

Accessibility is a very important quality in a good teacher. Availability shows interest on the professor's part, and this interest is usually returned by the students. —Wade Satterfield, Junior at Hendrix College.

No student mentioned knowledge of subject as a criterion for being a good teacher. This is probably because knowledge is assumed of anyone claiming to be a teacher. I was struck by the fact that nearly all these students held the same opinion: The ability to form positive relationships with students seems to be the most important characteristic of a good math teacher, even beyond that of giving interesting classes.

A Mathematical Time Line

1938

In 1938, the Polish statistician Jerzy Neyman (1894–1981) secured a position as a mathematics professor at the University of California. The head of the mathematics department was Griffith C. Evans (1887–1973). The following is an excerpt from the biography of Neyman.

> The week after Neyman's arrival, some twelve thousand students materialized on the campus, and almost immediately Evans asked him [Neyman] to assist with the registration of freshmen who wished to take mathematics courses.
>
> "So I was sitting at one desk on the right, Evans was sitting there, and so there was a crowd of students in front. And so he invited six, I remember—took their names, this and that, and then, 'Please come to the blackboard.' And that was a girl. All right. So she came. She didn't anticipate this. 'Please write down one-half plus one-third.' She looked at him and then wrote. 'Now write "equals" and then see what you get.' She looked at him indignantly and wrote one plus one, two plus three. Two fifths! Well. Evans shook his head. 'You go to the left. You will be in that section. Next please.' That was the way he sorted the students."
>
> "I can't believe it," I say.
>
> "True," Neyman insists. "True."
>
> —Constance Reid, *Neyman—From Life* (Springer-Verlag, 1982), page 160.

1943

Teaching at the university [Syracuse University] was pretty discouraging. There were many students who couldn't be talked out of $\frac{1}{2} + \frac{1}{3} = \frac{2}{5}$. —Paul R. Halmos, *I Want to be a Mathematician* (Springer-Verlag, 1985), pages 108–109.

1950

How is "one-fourth of one percent" written as a decimal?

    (A) $.0\frac{1}{4}$ (B) .025   (C) .0025   (D) .25

The correct answer (C) was selected by only 27% of 377 college freshmen in 1950. The freshmen were enrolled in a New York state teacher's college and their mean intelligence was tested and found to be slightly above the national average for college freshmen. Note that a score of about 25% would be expected if every student made a random guess. —Ben A. Seultz, *The Mathematics Teacher,* vol. 44, no. 1 (January 1951), page 13.

1979

A fourth of the nation's 17-year-old students cannot multiply 671 by 402 and get the right answer: 269,742. And the same multiplication problem baffles one-third of all thirteen-year-olds. —*Time,* "Education" (September 24, 1979). [This was based on an NAEP (National Assessment of Educational Progress) sampling of 71,000 U.S. students.]

1983

The 1983 report of the third NAEP by the National Commission on Excellence showed that less than 40% of American 17-year-olds could solve the two problems below.

1) Estimate the answer to 3.04×5.3.

    (A) 1.6   (B) 16   (C) 160   (D) 1600

Note: Only 37.6% chose the correct answer (B).

2) Find the distance from K to J. [A right triangle JKC was drawn with the right angle at C. The lengths of two sides were labeled: $JC = 400$ m and $KC = 300$ m.]

Only 39% chose the correct answer: 500 m.

1985

Using the letter S to represent the number of students (at this university) and the letter P to represent the number of professors, write an equation that summarizes the following sentence: "There are six times as many students as professors at this university."

The answer is obviously $S = 6P$. Trivial, no? The error rate for first-year engineering students at the University of Massachusetts is about 37%, with virtually all incorrect answers being of the form $P = 6S$. For students not majoring in science, mathematics, or engineering, the failure rate exceeds 50%. —Alan H. Schoenfeld, *Mathematical Problem Solving* (Academic Press, 1985), pages 64–65.

1992

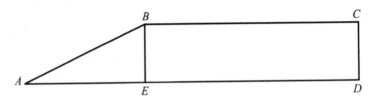

FIGURE 11

The area of rectangle $BCDE$ shown above is 60 square inches. If the length of AE is 10 inches and the length of ED is 15 inches, what is the area of trapezoid $ABCD$, in square inches?

The answer (80) was missed by 77% of the thousands of American high school seniors who took the test. The following question came from the same test:

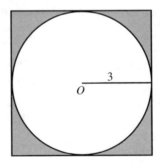

FIGURE 12

In the figure above, a circle with center O and radius of length 3 is inscribed in a square. What is the area of the shaded region?

A) 3.86   B) 7.73   C) 28.27   D) 32.86   E) 36.00

The answer is B) 7.73. Incredibly, 63% of the high school seniors who tried to answer this question chose the wrong answer. Source for both questions: Ina V.S. Mullis, et al., *NAEP 1992 Mathematics Report Card for the Nation and the States* (GPO, April 1993), page 737, Office of Educational Research and Improvement, U.S. Dept. of Education.

> [T]he percentage of students performing at the top levels on the national assessments is shockingly low, especially considering that the definition of excellence on these tests is exceedingly liberal. To earn a top score on the mathematics portion of the NAEP, for example, a high school junior need not demonstrate any skills beyond algebra. Yet, in 1992, as in 1978, only 7 percent of high school juniors scored in this category. —Laurence Steinberg, *Beyond the Classroom* (Simon & Schuster, 1996), pages 32, 33.

"Math class is tough." —Barbie Doll (1992)

Remarks about Learning Mathematics

The prolific science fiction writer Isaac Asimov took a Ph.D. in chemistry and taught chemistry as a university professor. He was also a member of Mensa and therefore had an IQ in the top 2% of people tested. Listen to what he said about his ability at math:

> After years of finding mathematics easy, I finally reached integral calculus and came up against a barrier. I realized that that was as far as I could go, and to this day I have never successfully gone beyond it in any but the most superficial way. —Isaac Asimov, *I Asimov* (Doubleday, 1994), page 91.

Most people do not possess, and do not easily acquire, the power of abstraction requisite for apprehending geometrical conceptions, and for keeping in mind the successive steps of a continuous argument. —G.A. Wentworth, *Plane and Solid Geometry,* rev. ed. (Ginn and Co., 1899), page iii.

THE QUESTION: Can significant improvements in a student's general critical thinking ability be achieved in a plane geometry class?

THE STUDY: Harry Lewis, *An Experiment in Developing Critical Thinking Through the Teaching of Plane Demonstrative Geometry* (New York University, 1950).

[THE CONCLUSION:] Geometry as ordinarily taught does little to improve general critical thinking ability. —"Research In Mathematics Education," edited by Dr. John Kinsella (New York University), *The Mathematics Teacher,* vol. 43, no. 8 (December 1950), pages 411, 412.

THE QUESTION: Does the study of Solid Geometry improve space perception abilities?

THE STUDY: Ernest R. Ranucci. *The Effect of the Study of Solid Geometry on Certain Aspects of Space Perception Abilities.* Ph.D. dissertation. (Teachers College, Columbia University, 1952).

[THE CONCLUSION:] For at least the past thirty years leaders in the field of the teaching of mathematics have claimed that the improvement in space perception abilities is one sound reason for the study of solid geometry. Dr. Ranucci's study was designed to test this hypothesis. The findings were startling: no statistically significant differences were found ... The claim that the study of solid geometry will improve space perception abilities ... has little statistical backing. —"Research In Mathematics Education," edited by Dr. John Kinsella (New York University), *The Mathematics Teacher,* vol. 45, no. 8 (December 1952), pages 621, 626.

It is clear that the chief end of mathematical study must be to make the students think. If mathematical teaching fails to do this, it fails altogether. The mere memorizing of a demonstration in geometry has about the same educational value as the memorization of a page from the city directory. And yet it must be admitted that a very large number of our pupils do study mathematics in just this way.

[T]he authors of practically all of our current textbooks lay all the emphasis on the formal logical side, to the almost complete exclusion of the psychological ... The textbook which takes due account of this psychological element is apparently still unwritten. —John Wesley Young (1911), *Lectures on Fundamental Concepts of Algebra and Geometry* (Macmillan, 1939), pages 4, 5.

Only professional mathematicians learn anything from proofs. Other people learn from explanations. I'm not sure that even mathematicians learn much from proofs in fields with which they are not familiar. —Ralph Boas, *Lion Hunting and Other Mathematical Pursuits* (Mathematical Association of America, 1995), page 234.

It does not necessarily follow, because children reproduce the language of reasoning, that they have reasoned. —A statement from an elementary school headmaster (c. 1912) Found in *American Mathematical Monthly,* vol. 78, no. 10 (December 1971), page 1081.

Information is very much easier to teach than thinking. . . . Skill in thinking is a broad skill like the skill in woodwork: knowing what to do, when to do it, how to do it, what tools to use, the consequences, what to take into consideration. It is much more than knowing the rules of logic or learning how to avoid logical errors. Skill in thinking has much to do with perception and with attention-directing . . . Whereas untrained pupils make an initial judgment and then generate only points that support that judgment, trained pupils are able to generate points that oppose their own view as well as those that support it. . . . Our obsession with logical error has not only failed to deal with the major causes of poor thinking but has prevented us from paying attention to those causes. —Edward de Bono, *Teaching Thinking* (Penguin, 1976), pages 34, 51–52, 66.

The old theory, which maintained that the study of mathematics, as an entity in itself, would promote the habits of logical thought in any other situation, is no longer held. . . . [T]raining in reasoning is rather specific so that reasoning cannot be done in any field without experience in that field. —Roy Dubisch, *The Teaching of Mathematics* (John Wiley & Sons, 1963), pages 9, 26.

It is my feeling that most people maintain the old theory, research notwithstanding.

The subject [of mathematics] has apparently no value beneficial to the development of mental power. Children who have learned algebra are not better thinkers than those who have not learned it. —Susanne K. Langer (Radcliffe College), *The Mathematics Teacher* (February 1966), page 158.

Actually, most mathematics courses do not teach reasoning of any kind. Students are so baffled by the material that they are obliged to memorize in order to pass examinations. —Morris Kline, *Why the Professor Can't Teach* (St. Martin's Press, 1977), page 128.

Summarizing the results of 75 studies on problem-solving strategies, E.G. Begle wrote the following:

A substantial amount of effort has gone into attempts to find out what strategies students use in attempting to solve mathematical problems. . . . No clear-cut directions for mathematics education are provided by the findings of these studies. In fact, there are enough indications that problem solving strategies are both problem- and student-specific often enough to suggest that hopes of finding one (or few) strategies which should be taught to all (or most) students are far too simplistic. —E.G. Begle, *Critical Variables in Mathematics Education* (Washington D.C.: Mathematical Association of America and National Council of Teachers of Mathematics, 1979), pages 145, 146.

"problem-solving skills." Work on the problem-solving abilities of specialists like doctors, chess players, and physicists has shown consistently that the ability to solve problems is critically dependent on deep, well-practiced knowledge within the special domain, and that these problem-solving abilities do not readily transfer from one domain to another. In short, there seems to exist no abstract, generalized, teachable ability to solve problems in a diversity of domains. For schools to spend time teaching a general skill that does not exist is clearly a waste of resources. —E.D. Hirsch, *The Schools We Need and Why We Don't Have Them* (Doubleday, 1996), page 264.

Why not just teach cooking and call it geometry? That way every student will be successful. And they can even eat their homework. —Joan Quill, Fairfax, Virginia, mathematics teacher (1990).

Joan Quill made that remark after hearing suggestions about changing the content of geometry—especially the omission of proofs—so that the weaker students would be more successful.

6
The Devil in the Classroom

Beware the devil in the classroom:

Teacher: "Mary, why don't you do your homework?"
Mary: "I did it for today, but I lost it."

Teacher: "George, why did you fail my test?"
George: "It was too hard, and there wasn't enough time to finish the test."

When students are asked questions about why they failed to complete an assignment or to study effectively, they feel confronted and will usually say the first thing that pops into their heads. In the past, if a student told me he had lost his homework, then I would talk to him about getting better organized. But, in fact, the student may have been well organized. He just may not have wanted to admit that he didn't do his homework. What students will say in their defense often isn't the case. I call this defensive lying, and I suspect that most adults and children respond with defensive lies without even thinking. I don't know how to correct the problem, but I do know how to work around it. The first step is to realize that placing a student on the defensive usually leads to disinformation for the teacher and bad feelings for the student. It is better to explore the causes with the student.

Teacher: "Mary, how can we get your homework grade up?"
Mary: "I don't know."

Teacher: "Do you have a place at home to do your work?"

Mary: "Yes, I work at my desk in my room."

Teacher: "Do you forget your assignments sometimes?"

Mary: "No, I usually remember them, but sometimes I don't understand how to do them."

Teacher: "If you start them in class, I can answer some of your questions for the first problems. Maybe that will help. What do you think?"

Mary: "I'll try."

With a smaller class, Mary could receive more assistance. But with the large classes that most of us teach, Mary must help herself. The teacher only diagnosed the problem for her. The reason I celebrate this kind of conversation is that it addresses a problem without making a bad situation worse. And that may be all that can be done. The trick of life is to do the best we can with what we have.

One time two students turned in identical computer programs. I talked to each one separately. The conversations were pretty much the same and went like this:

"Amy, why did you and Mark turn in the same program?"

(No response.)

"The two programs I received from you were identical except for the names and descriptions. I could hold them together in front of a window and every letter on one would line up under every letter of the other."

"We worked on it together."

"Did you type your program?"

"No (smiling), but I typed half of it."

"You know that you're supposed to do your own work. You're not being completely honest when you turn in work that is not all yours."

"I know, but I couldn't do the program and didn't want to lose credit."

"Amy, if you do things like this, people who know you well will eventually find out and have less respect for you. If someone you like comes to say, 'Oh, Amy is a nice enough person, but you can't always trust her; you can't always believe what she says,' that would be a terrible thing. When you turned in this program as your own work, you traded so much for so little. On reconsideration, don't you really think that you made a mistake here?"

"I guess. I just didn't want to get a zero."

"Here's what I'd like you to do for me. Write me the briefest one-paragraph essay—for my eyes only—explaining why you did this and what you now think you should have done instead. Give it to me by the end of the period."

The essay was the type of thing any student would write. Amy admitted it was a mistake, said she would never do it again, and apologized. I was lucky that she didn't find my words insincere or my manner patronizing. The next time I saw her I said, "Amy, I read your essay, and you have redeemed yourself in my eyes." I told Mark the same thing. Everybody won.

Accusations of cheating can make a student so defensive that he or she can't listen and will try to justify the actions as reasonable. I have seen students and parents try to deny charges or mitigate them, or turn them around so that the cheating is construed as students learning together. The result is that either nothing is done or the student is transferred to another teacher's class. But dwelling on this is counterproductive, because even if the teacher is vindicated, too much emotional damage is wrought through public confrontations. Embarrassment, criticism, and punishment are poor ways to change people, because they have too many unhealthy side effects. In my opinion, it is better to have students confront themselves.

I once had a talented geometry student named Karen who was caught up in being a teenager and perhaps had some personal problems that I didn't know about. She cut class many times and every time I turned her in, she would lie about where she was. When her counselor and I would confront her with proof, she would modify her story with another lie. We never did get her to admit that she cut class. She received an A the first quarter, but her grades went down after that. By the end of the year I could get her to work in class, but little more. She always had excuses for why her homework wasn't done. I would say, "Karen you are lucky that you have me as your teacher. I'm not going to give you a zero, because you have a good excuse. I'm going to let you turn your homework in tomorrow." The next day she would have another almost-believable excuse. She had a kind of genius for lying.

Both her counselor and I had several talks with her about how repeated lying would ultimately destroy the important relationships in her life. She listened politely. Later I asked her counselor if we had any effect. "Yes, we did. Karen went to the head of guidance and asked for a new counselor." Karen once asked me for a pass to the clinic and never returned to class or went to the clinic. The next time she asked for a pass, I told her, "The only way I will ever give you a pass to the clinic again will be if you show me an open wound." Despite my disapproval of her actions, Karen always remained polite to me. We never had words.

By the end of the last quarter, Karen had a failing grade and a major test to make up. She promised to come in eighth period or after school to make it up, but she never did. So I made her take it in parts during class. When she got certain problems wrong, I told her what the errors were and insisted she rework them. She caught on quickly and eventually got an A on the test. When I posted the class's final grades for the quarter, her grade was D+. On the last day of class I asked her to come to my desk. "Your grade is D+, Karen. You are so gifted that I am shocked to see this grade. I noticed that you didn't turn in the ten-point trigonometry packet. However, you did well on the trig test. So I suppose you mastered the material without doing all the assignments, and I know you did some of them in class. If I don't count the packet against you, your grade will jump up to a C+. That's still not a very good grade, but it's all that I feel I can do to improve your grade." She said, "Thank you, Mr. Stueben," and walked away. A lot of teachers would shake their heads over my actions. Let them disagree. Both the strong and the weak need to be looked after, and looked after in different ways.

AaBbCcDdEeFfGgHhIiJjKkLlMmNn

Part II

The Scrapbook

7
Mathematical Humor

Two mathematicians were having dinner. One was complaining: "The average person is a mathematical idiot. People cannot do arithmetic correctly, cannot balance a checkbook, cannot calculate a tip, cannot do percents, ..." The other mathematician disagreed: "You're exaggerating. People know all the math they need to know."

Later in the dinner the complainer went to the men's room. The other mathematician beckoned the waitress to his table and said, "The next time you come past our table, I am going to stop you and ask you a question. No matter what I say, I want you to answer by saying 'x squared.'" She agreed. When the other mathematician returned, his companion said, "I'm tired of your complaining. I'm going to stop the next person who passes our table and ask him or her an elementary calculus question, and I bet the person can solve it." Soon the waitress came by and he asked: "Excuse me, Miss, but can you tell me what the integral of $2x$ with respect to x is?" The waitress replied: "x squared." The mathematician said, "See!" His friend said, "Oh, ...I guess you were right." And the waitress said, "Plus a constant."[1]

For the purpose of a psychological experiment, a physics teacher, a businessman, and a pure mathematician were all asked to build the shortest possible fence around a small herd of resting cattle. The physics teacher was first. He took out a piece of graph paper and plotted the position of each cow.

---

[1] Thanks to my former computer science student Matt Blum (1992).

Next he introduced xy-axes and gave each cow a pair of coordinates. Then he determined the lines connecting all the points. Finally he constructed a fence based on his diagram. When he finished, he turned to the others and said, "I'm done. And since the interior region bounded by the line segments connecting the cattle-points is convex, it follows that the boundary is minimal. Q.E.D."

Then it was the businessman's turn. He began by calling the physicist an idiot and saying that it was no wonder young people entering the work force were useless with teachers like him. And if the physicist worked in his company, then he would fire him and fire the man who hired him. Finally he calmed down and went to work. First he secured a strong fence-pole near the cattle. Next he attached one end of a six-foot-high roll of wire fence to the pole and walked around the cows, slowly letting out the roll of wire fence until he came back to the post. Then he gave the roll to his assistant and told him to start pulling. The businessman ran around the outside of the fence kicking the cows, flailing his arms, and screaming at them to make them get up and move into the middle. All the while, he was yelling to his assistant: "Pull the fence tighter! Pull the fence tighter!" Finally, the cows were shoved so close together that they couldn't move, and the fence was wrapped around them so tightly that it was leaving marks on their hides. The businessman nailed the other end of the fence to the post, cut away the roll and said, "There, that is the shortest fence."

Finally it was the mathematician's turn. He had been solving differential equations in his head and had wandered away and had forgotten about the problem. So they had to find him and explain it to him all over again. Then he asked questions that nobody could understand like, "Am I allowed to assume a Euclidean metric?" Finally, the businessman threatened to hit him if he didn't get started. "Okay, okay, but I think the restraints are ill-defined," he said. Then he walked over to the roll of wire fence, cut off a small piece, wrapped it around himself, and said, "I'm on the outside."

A philosopher, a physicist, a biologist, and a mathematician were sitting outside in a street cafe. As they were conversing, they noticed two people walk up to a parked van, open it, and get in. After a few minutes three people got out and walked away.

"If we assume the van was empty to start with, how is that possible?" asked the philosopher.

"Clearly, our assumptions are inaccurate," replied the physicist.

"Or they reproduced," chuckled the biologist.

"If one more person gets in, then the van will be empty," said the mathematician, seriously.

The businessman says: "My business can control the economy."
The economist says: "The forces of the economy will control your business."
The biologist says: "I study the principles of life."
The psychologist says: "You are controlled by the principles of life."
The physicist says: "The universe is a model of my equations."
The engineer says: "Your equations are a model of the universe."
The philosopher says: "Truth cannot be distinguished from illusion."
The logician says: "Truth is defined by well-formed formulas."
The mathematician says: "I don't care."

Many of these jokes make fun of mathematicians by stereotyping them as absent-minded professors cut off from reality. That, of course, is the humor; they are extreme to the point of being ridiculous. This last joke, however, is not only a joke—it's true. It is the job of a scientist to apply mathematics. It is the job of the mathematician to generalize concepts, to determine limits, to establish logical foundations, and to sharpen the tools. The pure mathematician must live in an ivory castle, because that is where his work is. As a human being, the mathematician may care that mathematics is applied to solve human problems, but as a mathematician, his primary interest is in the abstract properties of the tools.

Seminar Attendee: "Sir, I can think of *two* counterexamples to your last statement."
Seminar Speaker: "That's all right. I have *three* proofs that it is true."

Science Humor

Science Club Notice: A seminar on time travel will be held on Thursday last week. The invited speaker, Prof. Blake DeKalb (an expert in palindromic numbers) from Northern Illinois University, was a member of the Bermuda Triangle Expeditionary Force (2003–1989). The subject will be "How to Buy a Time Travel Machine." Synopsis: Time travel machines are not yet available, and when they become available they will be very expensive. It is suggested that you buy one at any cost. Then go back in time and give it to yourself.

Then go forward in time and tell yourself not to buy the machine, because you already own it.

Fredric Brown's "First Time Machine" opens with a Dr. Granger exhibiting his time machine to three friends. One of them uses the device to go back sixty years and kill his hated grandfather when the man was a youth. The story ends sixty years later with Dr. Granger showing his time machine to two friends. —Martin Gardner, *Time Travel and Other Mathematical Bewilderments* (Freeman, 1988), page 3.

gravitation, n. The tendency of all bodies to approach one another with a strength proportional to the quantity of matter they contain— the quantity of matter they contain being ascertained by the strength of their tendency to approach one another. This is a lovely and edifying illustration of how science, having made A the proof of B, makes B the proof of A. —Ambrose Bierce [1842–1913?], *The Devil's Dictionary* (Neale Pub. Co., 1911; Dover reprint, 1958).

Thomas Carlyle (1795–1881) was a Scottish essayist and historian. He is considered to be one of the great sages of his era. Early in his career, he taught mathematics and translated Legendre's *Geometry* from French to English. In 1818 he left the profession of teaching mathematics and later remarked, "Teaching school is but another word for sure and not very slow destruction."[2]

Carlyle was only eleven months old and had never spoken a word, when hearing another child in the household cry, he sat up and said "What ails wee Jock?"[3]

"The one feature of his [Einstein's] childhood about which there appears no doubt is the lateness with which he learned to speak. Even at the age of nine he was not fluent, ...His parents feared that he might be subnormal.... " —Ronald W. Clark, *Einstein: The Life and Times* (Avon Books, 1984), page 27.

---

[2] The *Mathematics Teacher* vol. LIX, #8 (December 1966), pages 757, 770.

[3] John W. Gardner, *Excellence,* revised ed. (Norton, 1984), page 74.

Supposedly Albert Einstein never talked until he was five years old. His parents were worried that he may have been retarded, but doctors couldn't find anything wrong with him—except that he wouldn't talk. Then one night at supper, young Albert said, "The soup is too hot." His parents were ecstatic. Albert was not retarded; he could talk. After they calmed down, they were curious and asked him why he had never uttered a word before. He replied, "Because before now everything was satisfactory."

"The fact that I neglected mathematics to a certain extent had its cause not merely in my stronger interest in the natural sciences than in mathematics, but also in the fact that my intuition was not strong enough in the field of mathematics in order to differentiate clearly the fundamentally important, that which is really basic, from the rest of the more or less dispensable erudition." —Albert Einstein, "Autobiographical Notes," in Paul A. Schilpp (ed.), *Albert Einstein: Philosopher-Scientist* (Tudor Publishing, 1951), page 15.

Albert Einstein (1879–1955) and the French mathematician Henri Poincaré (1854–1912) were only known to have met once (at a conference in 1911). There is a legend that Poincaré discovered the special theory of relativity before Einstein.[4] In 1990, David Singmaster told me that the following conversation had occurred between Einstein and Poincaré:

Einstein: "You know, Henri, I once studied mathematics, but I gave it up for physics."

Poincaré: "Oh, really, Albert, why is that?"

Einstein: "Because although I could tell the true statements from the false, I just couldn't tell which facts were the important ones."

Poincaré: "That is very interesting, Albert, because, I originally studied physics, but left the field for mathematics."

Einstein: "Really? Why?"

Poincaré: "Because I couldn't tell which of the important facts were true."

The word *schedule* is pronounced shed·yoo·al by the British. In the 1930s, Albert Einstein and Winston Churchill were having a conversation.

---

4 Jeremy J. Gray, *The Mathematical Intelligencer*, vol. 17, no. 1 (winter 1995), pages 65–67, 75.

Near the end of the conversation, Einstein said, "I'll have to check my shed-yoo-al." Churchill, who had not expected the correct English pronunciation from a foreigner, replied: "Professor Einstein, I want to compliment you on your particularly accurate pronunciation of our language. Do you mind telling me where you learned to speak English?" To which Einstein replied: "Ya, sure. In shool."

We [Einstein and Ernst Straus] had finished the preparation of a paper and were looking for a paper clip. After opening a lot of drawers we finally found one which turned out to be too badly bent for use. So we were looking for a tool to straighten it. Opening a lot more drawers we came upon a whole box of unused paper clips, Einstein immediately started to shape one of them into a tool to straighten the bent one. When asked what he was doing, he said, "Once I am set on a goal, it becomes difficult to deflect me." —Ernst G. Straus, "Memoir," in A.P. French (ed.), *Einstein: A Centenary Volume* (Harvard University Press, 1979), page 31.

A scientist had two large jars before him on the laboratory table. The jar on his left contained a hundred fleas; the jar on his right was empty. The scientist carefully lifted a flea from the jar on the left, placed the flea on the table between the two jars, stepped back and in a loud voice said, "Jump." The flea jumped and was put in the jar on the right. A second flea was carefully lifted from the jar on the left and placed on the table between the two jars. Again the scientist stepped back and in a loud voice said, "Jump." The flea jumped and was put in the jar on the right. In the same manner, the scientist treated each of the hundred fleas in the jar on the left, and each flea jumped as ordered. The two jars were then interchanged and the experiment continued with a slight difference. This time the scientist carefully lifted a flea from the jar on the left, yanked off its hind legs, placed the flea on the table between the jars, stepped back and in a loud voice said, "Jump." The flea did not jump, and was put in the jar on the right. A second flea was carefully lifted from the jar on the left, its hind legs yanked off, and then placed on the table between the two jars. Again the scientist stepped back and in a loud voice said, "Jump." The flea did not jump, and was put in the jar on the right. In this manner, the scientist treated each of the hundred fleas in the jar on the left, and in no case did a flea jump when ordered. So the scientist recorded in his notebook the following induction: "A flea, if its hind legs are yanked off, cannot hear."[5]

Theodore von Karman (1881–1963), the great Hungarian aeronautical engineer, once told the following story explaining why he chose to study engineering in Germany rather than Hungary. When he was in high school, his physics teacher often praised German engineering. One day von Karman said to his teacher, in front of the class, that he didn't think German engineers could be any better than other engineers, because the same principles of engineering were taught everywhere. The class became silent, and the teacher stared at von Karman for a while.

Then the teacher said, "Sit down and listen to me. I will tell you a story. In the French Revolution there were three foreign engineers trapped in Paris: an Englishman, an Italian, and a German. Since they had been hired by the aristocracy, they were to be executed with the aristocracy."

"The three men were marched to the guillotine platform in front of a cheering crowd. The Englishman was first, but he had a request. He wished to lie on his back so that his last vision would be of the heavens. The executioner acquiesced, and he was placed on his back. The guillotine blade was slowly raised to the top, where it made a clicking sound just before it dropped. The Englishman cried out, 'My God! My God!' The blade came rushing down, and a miracle occurred. It stopped five inches above his throat. The crowd gave a great roar. 'It is a sign from God! He is innocent! He must be freed!' So they untied him and led him into the crowd."

"The next to be executed was the Italian. He also wished to lie on his back so that his last vision would be of heaven. The guillotine blade was slowly raised to the top, where it made a clicking sound just before it dropped. The Italian screamed, 'My God! My God!' The blade came rushing down, and again a miracle occurred. It stopped just four inches above his throat. The crowd again gave a great roar. 'He is innocent! He must be freed!' So they untied him and led him into the crowd."

"Then it was the German's turn. He wished to lie on his back too. The guillotine blade was slowly raised to the top, where it made a clicking sound just before it dropped. Suddenly the German cried out, 'Stop, I see it! The rope is slipping off the left pulley.'"

---

5 Howard Eves, *In Mathematical Circles Quadrants I and II* (Prindle, Weber & Schmidt, 1969), page 6. Two people have told me that they recall learning in biology classes that some insects hear vibrations with the hair on their legs. In that case, the scientist's conclusion is correct!

"It is not knowledge that makes great engineers—it is their passion for its applications."[6]

Mathematical Wordplay

René Descartes and his wife were hosting a late night party. The main course was small shrimp which had been attractively laid out on a dinner table. Dinner was to be served at 1 o'clock in the morning. Around midnight, Mrs. Descartes took René aside and told him to stand near the shrimp and keep the guests from snacking on them during the next hour or there would be no main course left for dinner. And he was further instructed to be tactful about this with the guests.

As René stood watch over the shrimp, a guest approached to engage him in conversation. During the conversation the guest reached over, picked up a shrimp, and popped it into his mouth. Descartes, noticed this, pulled out his note pad, wrote a brief note and passed it to the guest. The startled guest opened the note and read: "I think they're for 1 A.M."[7]

René Descartes was sitting outside of a Parisian cafe when a waiter approached him and asked, "More coffee, sir?" Startled from his musing at the moment, Descartes looked up and replied, "I think not!" and then disappeared.[8]

Desmond MacHale once wrote that no mathematical result is worse than no mathematical result. Get it? No *particular result* is so bad as to be worse than not getting a result at all. Well, maybe not. The world might be a better place if important research were not buried in a lot of trivial research. MacHale reports that John Conway once suggested a good test of the value of a paper is

---

[6] A version of this joke appeared in Betsy Devine and Joel Cohen, *Absolute Zero Gravity* (Simon & Schuster, 1992), pages 54–55.

[7] Thanks to Gretchen Goswik, 1987. Gretchen Goswik's joke, like so many others in this collection was told to me only once. I now dive for a pen the moment I hear anything interesting in mathematics.

[8] Unfortunately, the joke is not correct. If thinking implies existence, then it does *not* follow that no thinking implies non-existence. Descartes has been criticized for his famous statement. More accurate, some have claimed, would have been *I think, therefore my thoughts exist.* Ambrose Bierce preferred *I think that I think, therefore I think I am.*

to ask the author if he or she would consider never publishing it for £1000. If there is any hesitation, the paper should not be published. Morris Kline suggested preceding every published paper with a note as to its value—a suggestion that was never popular among editors and authors.

Mathematical Anagrams

1.	a decimal point	=	I'm a dot in place
2.	decimal point	=	I'm a pencil dot
3.	logarithm	=	algorithm
4.	a number line	=	innumerable
5.	integral calculus	=	calculating rules
6.	algebra	=	a garble
7.	calculation	=	I call a count
8.	higher mathematics	=	ahh! arithmetic gems
9.	inconsistent	=	n is, n is not, etc.
10.	negation	=	get a "no" in
11.	pocket calculators	=	clack! total up score
12.	the answer	=	wasn't here
13.	school master	=	the classroom
14.	listen	=	silent
15.	committees	=	cost me time
16.	incomprehensible	=	problem in Chinese

Left-handed Definitions

calculator. An instrument for extending guesswork to the 12th place.

positive. Mistaken at the top of one's voice —Ambrose Bierce (1911).

proof. One-half of one percent alcohol. (R. Graham, D. Knuth, O. Patashnik (1989)).

logic. The art of going wrong with confidence.

Q.E.D. 1) quod erat demonstrandum (= which was to be demonstrated). 2) quite easily done. 3) quite erroneously done. 4) question every deduction. 5) quick, an English dictionary. 6) quid erat Deus (God what was it?)

Humorous Proofs

Theorem 1.   A sheet of writing paper is a lazy dog [1].

Proof.   A sheet of paper is an ink-lined plane. An inclined plane is a slope up. A slow pup is a lazy dog. Therefore, a sheet of writing paper is a lazy dog.[9]

Theorem 2.   A peanut butter sandwich is better than eternal happiness.

Proof.   A peanut butter sandwich is better than nothing. But nothing is better than eternal happiness. Therefore, a peanut butter sandwich is better than eternal happiness.[10]

Theorem 3.   A crocodile is longer than it is wide.

Proof.   A crocodile is long on the top and the bottom, but it is green only on the top; consequently, a crocodile is longer than it is green. A crocodile is green along both its length and width, but it is wide only along its width; consequently, a crocodile is greener than it is wide. Therefore, a crocodile is longer than it is wide. Q.E.D.

Theorem 4.   Every horse has an infinite number of legs.

Proof.   Horses have an even number of legs. Behind they have two legs, and in front they have fore legs. This makes six legs, which is certainly an odd number of legs for a horse. But the only number that is both odd and even is infinity. Therefore, horses have an infinite number of legs.[11]

Theorem 5.   Napoleon was a poor general.

Proof.   Most men have an even number of arms. Napoleon was warned that Wellington would meet him at Waterloo. To be forewarned is to be forearmed. But four arms is certainly an odd number of arms for a man. The only number that is both even and odd is infinity. Therefore, Napoleon had an infinite

---

[9]   Harry Kurus, in John G. Fuller, *Games for Insomniacs* (Doubleday, 1966), pages 73–74.

[10]   A variation of this verbal fallacy can be found in Henry Coppee, *Elements of Logic* (J. H. Butler and Co., 1881), page 159.

[11]   Joel Cohen, *The Worm Runner's Digest,* vol. III, no. 3 (December 1961).

number of arms in his battle against Wellington. Since Napoleon still lost the battle, he must have been a very poor general indeed.[12]

Theorem 6. If $\frac{1}{0} = \infty$, then $\frac{1}{\infty} = 0$.

Proof (by rotation). Given $\frac{1}{0} = \infty$, rotate both sides 90° counterclockwise and obtain $-10 = 8$. Subtract 8 from both sides: $-18 = 0$. Finally, rotate both sides 90° in the reverse direction: $\frac{1}{\infty} = 0$.

Theorem 7. Death comes to no man.

Proof. As is well known and celebrated in legend and song, when we approach death, our whole life flashes in front of us. This short review—if it is to be complete—must also include the moment we approached death and the flashback of our life. But this second flash must by the necessity of completeness include another flash of life. And that flashback must include still another and another, etc., etc. Hence, although we may approach death, all eternity is not enough time for us to reach it. —This is known as "Leinbach's Proof" from *Flight into Darkness* by Arthur Schnitzler.

Theorem 8. All governments are unjust.

Proof. To establish the truth of this proposition we need only show it true for an arbitrary government. But by definition, a government that is arbitrary is unjust.

Theorem 9. Christmas = Halloween = Thanksgiving (at least for assembly language programmers).

Proof. By definition, Christmas = Dec. 25; Halloween = Oct. 31; Thanksgiving = Nov. 27, sometimes. Again by definition,

```
Dec 25 is 25 base 10 or (2 x 10) + (5 x 1) = 25.
Oct 31 is 31 base  8 or (3 x  8) + (1 x 1) = 25.
Nov 27 is 27 base  9 or (2 x  9) + (7 x 1) = 25.¹³
```

---

[12] This is a variation of Joel Cohen's proof in *A Stress Analysis of a Strapless Evening Gown*, Robert A. Baker, ed. (Prentice-hall, 1963), page 89.

[13] Solomon Golomb and John Friedlein, in Martin Gardner's *Mathematical Magic Show* (Alfred A. Knopf, 1977), pages 72 and 80.

Theorem 10.   The positive integer 10 is even.

Proof.
 (1) $9 - 6 = 3$.
 (2) IX $-$ SIX $= -$S.
 (3) $\therefore 3 = -$S.
 (4) $10 = 7 + 3$.
 (5) $10 = 7 + (-$S$)$.
 (6) $10 =$ SEVEN $+ (-$S$)$.
 (7) $\therefore 10 =$ EVEN.[14]

Theorem 11. $7 \times 13 = 28$.

Proof.

```
Check #1: 13      Check #2:    13
          × 7               7) 28
          21               -  7
          + 7                 21
          28                -21
                              0

Check #3: 13
          13
          13
          13
          13
          13
         +13
          28  =  (3+3+3+3+3+3+3)  +  (1+1+1+1+1+1+1)
```

 —From the 1941 Universal Pictures movie "In the Navy" starring Bud Abbott and Lou Costello.[15]

----

[14] R. M. Abraham, *Winter Night's Entertainments* (1935; reissued as *Tricks and Entertainments* by Dover, 1964), page 12.

[15] The chubby Lou Costello and his tall straight man Bud Abbott were a comedy team from 1937 to 1956. They made more than thirty movies. The actual comedy dialog accompanying this proof may be found in Richard J. Anobile (ed.), *Who's on First* (Avon Books, 1972), pages 129–139. Anobile noted that the Abbott and Costello movies had never been transferred from the nitrate stock and were slowly deteriorating. It is not clear how many, if any, of the Abbott and Costello movies still exist today.

Question 1: A man runs t times round a circular track whose radius is n miles. He drinks s quarts of beer for every mile that he runs. Prove that he will drink exactly one quart of beer!

Answer 1: Since the circumference of a circle is $2\,pi$ times the radius n, we need only multiply this by t times s to obtain our answer: $2\,pi\,nts =$ 1 quart.[16]

Question 2: There is a butcher who is 60 years old, 60 inches tall, and has a 60-inch waist. What does he weigh? Big Hint: The significant verb in this problem is transitive. See note [2].

Question 3: A bag contains exactly two marbles, each of which is either white or black. What are the colors of the two marbles?

Answer 3: One is white and the other is black.

Proof. From the given information we know that if a marble is randomly chosen there is a 50% chance that it will be black. Now if both marbles were black, the chance would be 100%. If both marbles were white, then the chance would be 0%. Consequently, the marbles must be of different colors.[17]

Notes

1. I sometimes introduce this argument when the transitive property is discussed in class. The word transitive was probably chosen for verbs that take objects because these verbs transmit an action to an object. The term transitive was probably chosen to describe relations like equality because, when three or more relationships are linked together, the relationship seems to pass across or through the chain from the first to the last element.

2. Answer 2: The butcher weighs meat.

[16] Henry Dudeney (1857–1931), *Modern Puzzles* (1926), reprinted in *536 Puzzles and Curious Problems,* edited by Martin Gardner (Charles Scribner's Sons, 1967), pages 22 and 239.

[17] Lewis Carroll (1895), reprinted in *Pillow Problems and a Tangled Tale* (Dover, 1958), pages 18, 27, and 109. However, the proof given here is different from Lewis Carroll's.

8
Mnemonics

[I]solated items are laboriously acquired and easily forgotten. Any sort of connection that unites the facts simply, naturally, and durably, is welcome here. The system need not be founded on logic, it must only be designed to aid the memory effectively.
—George Pólya, *How To Solve It* (Princeton, 1945), page 218.

In 1492 Columbus sailed the ocean blue.
Or was it
In 1493 Columbus sailed the deep blue sea?
—Anonymous

Mnemonics are memory aids. The word "mnemonic" comes from Mnemosyne, the Greek goddess of memory and mother of the muses. Mnemonics can be both useful and a form of recreational word-play. For a mnemonic to be useful, it must be wisely constructed. The author of the following parody tried to warn teachers from carrying mnemonics too far.

What color are fire engines? Let me see. Before fire engines, men went to fires on foot. A foot is twelve inches. Twelve inches is a ruler. A ruler of the British Empire was Queen Mary. The Queen Mary is a ship. Ships sail the oceans. Across the ocean is Russia. A Russian is a red. Hence, fire engines are red. Such a simple method to use for remembering that fire engines—at least some of them—are red! ... Don't suggest mnemonic devices unless they are good ones. —Charles W. Mathews, *School Science and Mathematics,* vol. 49 (1949), pages 491 and 492.

Many teachers purposely avoid the use of mnemonics, because a person who uses a mnemonic has one more thing to remember, and mnemonics emphasize form over content. In 1967, Gerald R. Miller of the University of Texas at El Paso did a study on the effectiveness of mnemonic devices.[1] Working with 44 high school students, Miller gave them various types of mnemonics to help them in the study of history. Two conclusions were drawn: a) The use of mnemonic devices led to a marked improvement in test scores; and b) the students who used mnemonics showed no significant improvement in long term retention. In other words, a mnemonic device is effective only in memorizing for tests. Another disadvantage, according to John Collings Squire, is that mnemonics may return to haunt a user throughout his life.

> These mnemonic rhymes begin as one's servant and end by being one's master; how many of us find the wretched things ringing in our heads whenever we have to remember something, long after our common faculties have put us in full possession of the knowledge. The deuce of a mnemonic is that once acquired it can never be entirely lost. Either it comes grinding into your head whenever it has a chance or else its phantom haunts one until the rhyme is completely recalled. —John Collings Squire (pseud. Solomon Eagle), *Books in General* (1921).

A mnemonic, like any crutch, can sometimes lead to embarrassment. In the October 1955 issue of the *Scientific American,* George Gamow gave 3.14158 as a value of pi. Readers wrote in to point out an error in the last digit. Gamow replied that, as a small boy, he had learned a rhyme for pi in French. His faulty spelling was responsible for the error. Gamow then proceeded to give his mnemonic in "corrected" form. Again readers wrote to complain that he was still wrong. Most likely, Gamow had memorized an incorrect version as a child. In his autobiography he recalled:

> ...after a few readings of a verse, even a rather long one, I remember it for an indefinite period of time. Several years ago I visited some Russian friends in San Diego and made a bet that I could quote Russian verses from memory for at least one hour, nonstop. I won the bet, quoting verses for an hour and a half, and stopped not because I ran out of verses but because the audience got tired. The next day laryngitis forced me to cancel my lecture. —George Gamow, *My World Line* (Viking Press, 1970), page 20.

---

[1] Gerald R. Miller, *An Evaluation of the Effectiveness of Mnemonic Devices as Aids to Study* (University of Texas at El Paso, 1967).

Here is the mnemonic that Gamow thought he had memorized:

Que j'aime a faire apprendre un nombre utile aux sages!
Immortel Archimede, antique ingenieur,
Aui de ton jugement peut sonder la valuer?
Pour moi ton probleme eut de pareils avantages.
—Edouard Lucas, *Recreations Mathematiques,* vol. II
(Paris, 1896), page 155.

The author, François-Edouard Anatole Lucas (1842–1891), was a French professor of mathematics at the Lycée Saint Louis and creator of the famous Tower of Hanoi puzzle. He has been called the Martin Gardner of his time. His books on recreational mathematics are widely quoted in number theory, but have never been translated into English. Here is the literal translation of his rhyme:

How I would like to express a number useful to the wise.
Immortal Archimedes, ancient engineer,
Who can appreciate the value of your judgment?
For me your problem had equal advantages.

The last line probably refers to the fact that Lucas, like Archimedes, also hunted for fractional approximations to pi. Archimedes found $\frac{22}{7}$ around 240 B.C.; Lucas gave the more accurate $\frac{13\sqrt{146}}{50}$ in 1896—*Recreations Mathematiques,* vol. II, page 236. Lucas's rhyme has had considerable effect in motivating English rhymes. On August 5, 1905, the London literary magazine *The Academy* quoted Lucas' quatrain and asked its readers for English equivalents. The British scientific magazine *Nature* reprinted it that year, as did the American *Literary Digest,* and the *American Mathematical Monthly* in 1906. In 1914, the *Scientific American Supplement* also challenged its readers with Lucas' mnemonic. The results are provided in the next section.

Mnemonics for Pi

Pi = 3.14159 26535 89793 23846 26433 83279 502. . .

Pi seems to be the most popular number to mnemonicize. The set collected here was drawn from many out-of-print books and back issues of mathematical and literary journals. There are probably many more in the literature that I haven't found. In most of the following items, the number of characters in each word represents a digit of pi. Occasionally a 10-, 11-, 12- or 14-digit word is used to convey two digits of pi. Sometimes the word pi (2 characters) is used, and sometimes the symbol π (1 character) is used to make the mnemonic correct.

3.1416

May I draw a circle?

> —Author unknown, in Dmitri Borgmann,
> *Language on Vacation* (Charles Scribner's
> Sons, 1965), page 25.

3.1416

Yes, I have a number.

> —H. E. Licks, *Recreations in Mathematics*
> (Van Nostrand, 1917), page 50.

3.14159

Wow, I made a great discovery.

> —Author unknown.

3.14159 2

Now I need a verse recalling pi.

> —Irene Fischer/Dunstan Hayden, *Geometry*
> (Allyn & Bacon, 1965), page 307.

$113\overline{)355} \Rightarrow 3.141592\ldots$ In this figure, the first three odd digits are each repeated twice in succession. This is one of the easiest ways to remember pi to six decimal places.

3.14159 265

Yes, I know I shall recollect my number right.

> —Wallace Lee, *Math Miracles* (Chapel Hill,
> North Carolina, 1960), page 9.

3.14159 265

How I wish I could enumerate pi easily today.

> —Author unknown.

3.14159 265

May I draw a round enclosure as circle known?

> —Dmitri Borgmann, *Language on Vacation*
> (Charles Scribner's Sons, 1965), page 251.

3.14159 2653

$$\frac{\text{Calculator will get fair accuracy}}{\text{but not to } \pi \text{ exact}} = \frac{104348}{33215}$$

3.14159 26535 8

$$\frac{\text{Dividing top lot through (a nightmare)}}{\text{By number below, you approach } pi} = \frac{833719}{265381}$$

> —W. Hope-Jones, *Mathematical Gazette,* vol.
> 23, no. 255 (July 1939), page 284.

3.14159 2654

May I have a large container of coffee right now?

—Author unknown.

3.14159 26535

May I have a large container of coffee—sugar and cream?

—Author unknown.

3.14159 26535

If you can remember 3.14159, the next five places may be recovered by adding to the decimal part the terms of the series 1, 2, 4, 8, 16 neglecting all carrying.

$$
\begin{array}{r}
3.1415\ 9 \\
124816 \\
\hline
2653\ 5
\end{array}
$$

—S. A. Saunder, *Mathematical Gazette,* vol. 4 (October 1907), page 131.

3.14159 26536

But I must a while endeavour

To reckon right the ratios.

—Author unknown, *Mathematical Gazette,* vol. 10 (October 1921), page 323. [Note the British spelling of endeavor.]

3.14159 26535 8 (inspired by Lucas)

Sir! I send a rhyme excelling

In sacred truth and rigid spelling.

        [or]

In mystic force and magic spelling.

—F.R.S. (possibly a Fellow of the Royal Society who wished to remain anonymous), *Nature,* vol. 72, no. 1875 (October 5, 1905), page 558.

3.14159 26535 89

Now I live a drear existence in ragged suits

And cruel taxation suffering.

—Author unknown, in *The Dark Horse,* the staff magazine of Lloyds Bank (1951).

3.14159 26535 89793 23846
How I wish I could recollect pi
Eureka! cried the great inventor.
Christmas Pudding, Christmas Pie
Is the problem's very center.

> —Author unknown, in Alan D. Baddeley,
> *The Psychology of Memory* (Basic Books,
> 1976), page 349.

3.14159 26535 89793 23846
Now I sing a silly roundelay
Of radical roots, and utter, "Lackaday!
Euclidean results imperfect are, my boy ...
Mnemonic arts employ!"

> —Willard R. Espy, *An Almanac of Words at
> Play* (Clarkson N. Potter, 1975), page 44.

3.14159 26535 89793 2384 (inspired by Lucas)
Now, I have a score notations
    Of digits large and small,
Teaching diameter's precise relations,
    And we can remember 'tall.

> —G. E. Gude (Kansas City, Missouri), in the
> *Scientific American Supplement,* vol. 77,
> no. 1994 (March 21, 1914), page 190.

Several pi-rhymes refer to Archimedes who produced the best approximation of his time: $1\frac{10}{70} < \pi < 3\frac{11}{70}$.

3.14159 26535 897
How I wish I could recollect of circle round
the exact relation Archimede unwound!

> —Author unknown, *Proceedings of the
> Edinburgh Mathematical Society,* vol. III
> (May 1885), page 106.

Unfortunately the author dropped the *s* off of *Archimedes* (poetic license). Below is the world's best loved mnemonic for pi:

3.14159 26535 8979
How I want a drink,
Alcoholic of course,
After the heavy chapters
Involving quantum mechanics.

> —Sir James Jeans (pre-1932).

My reference is to a 1932 issue of the *American Mathematical Monthly,* where it was credited to Jeans from an unnamed source. I have never been able to track down the original source. A clever 9-digit extension is *All of thy geometry, Herr Planck, is fairly hard.* The German physicist Max Planck (1858–1947) originated and developed quantum theory.

Mathematicians will prefer this variation:

3.14159 26535 8979
God! I need a drink—
Alcoholic of course—
After all those lectures
Involving radical equations.

The next rhyme is a bright variation of a larger mnemonic which is reproduced later:

3.14159 26535 89793 23846 26
See, I have a rhyme assisting
My feeble brain,
Its tasks ofttimes resisting.
Further mnemonics for pi
Can maintain more digits in supply.
                              —Author unknown

3.14159 26535 89793 23846 264 (inspired by Lucas)
Now I know a spell unfailing,
An artful charm, for tasks availing,
Intricate results entailing.
Not in too exacting mood,
(Poetry is pretty good), ...
                —Author unknown, *Nature,* vol. 72, no. 1878
                (October 26, 1905), page 63.

The greatest mental calculator of all time was Johann Martin Zacharias Dase (sometimes Dahse), a native of Hamburg. According to W. W. Rouse Ball in his *Mathematical Recreations and Essays,* Dase was once asked to multiply mentally 79,532,853 by 93,758,479. He gave the 16-digit answer in 54 seconds. Asked how many letters were in a specific line of print, he instantly gave the correct answer of 63. When he was 16 years old, he calculated the decimal expansion of pi to 205 places in two months (the last five places were subsequently found to be in error). Ball remarks that Dase struck all observers as dull. Except for geometry and his native German language, he remained ignorant to the end of his days. But, like many others, Dase also felt inspired to compose a mnemonic for pi.

3.14159 26535 89793 23846 264

Wie? O! Dies

O Macht ernstlich so vielen viele Muh?

Lernt immerhin Junglinge, leichte Verselein

Wie so zum Beispiel dies möchte zu merken sein!

> —Johann Dase (1824–1861), *Wiskundig Tijdschfift,*
> vol. III (1905–1906), page 65; and *The Mathemat-*
> *ical Gazette,* vol. 4, no. 65 (July 1907), page 103.

3.14159 26535 89793 23846 26433 8

For circumscribing a round enclosure or circle every man

might remember ingenious numbers measuring one by one

diameter into circle or circle upon its own diameter. . . .

> —Author unknown, *Mathematical Gazette,* vol.
> 4, no. 65 (July 1907), page 103. [Notice the
> clever use of a 14-letter word in this rhyme.]

3.14159 26535 89793 23846 26433 83279 (inspired by Lucas)

See, I have a rhyme assisting

My feeble brain its tasks sometime resisting,

Efforts laborious can by its witchery

Grow easier, so hidden here are

The decimals all of circle's periphery.

> —L. R. Stokelbach (Detroit, Michigan), *The*
> *Scientific American Supplement,* no. 1994
> (March 21, 1914), page 190.

3.14159 26535 89793 23846 26433 83279 (inspired by Lucas)

Now I—even I—would celebrate

In rhymes unapt, the great

Immortal Syracusan,* rivaled nevermore

Who in his wondrous lore,

Passed on before,

Left men his guidance how to circles mensurate.**

> —Adam C. Orr, *Literary Digest,* vol. 32, no. 3
> (January 20, 1906), page 84. [* Archimedes;
> ** Compare as ratios]

A variation of Orr's mnemonic is as follows:

3.14159 26535 8793 23846 26433 83279

Now I—even I—would celebrate in rhymes

Inept the great immortal Syracusan

Rivaled nevermore, who by his wondrous love

Untold us before, made the way straight

How to circles mensurate.

—Author unknown.

In the August 12, 1905, issue of *The Academy,* a popular London literary magazine, a correspondent gave the criteria by which he felt a pi mnemonic should be judged:

> They are easy to construct, but the problem is how to obtain a mnemonic which shall contain as complete a reference as possible to the point in question: (1) That pi is the ratio of the circle to its diameter, (2) that the series is infinite, and (3) that the words translated into the number of their letters, give the value of pi.

I would give different criteria: (1) that they should say something entertaining or worth knowing and (2) that the mnemonic should rhyme. These two criteria are hard to satisfy.

3.14159 26535 89793 23846 26433 83279 (inspired by Lucas)

Let π mean a ratio (circulars to 'meters)

Whose end beats straight computers,

    staunch computers,

And, to ten decimals plus twenty, pi arises

From our pet verselet all in numeral disguises.

—W. Renton, The Academy, vol. 69, no. 1735
(August 12, 1905), page 796. [Note the use
of both "π"and "pi".]

3.14159 26535 89793 23846 26433 83279

Now I will a rhyme construct,

By chosen words the young instruct,

Cunningly devised endeavors,

Con it and remember ever

Widths in circle here you see

Sketched out in strange obscurity—.

—Author unknown, The Dark Horse, the staff
magazine of Lloyd's Bank (1951). [con =
to learn or commit to memory]

3.14159 26535 89793 23846 26433 83279 5

Sir: I wish I could recapture my memory about Sir

Jeans' diabolic mnemonics! However, invention

now of any reliable easy phrase is beyond what shy

and fumbling aid my present intellect gives.

—Bill Powers (Chicago), in Willey Ley,
The Borders of Mathematics (Pyramid
Publications, 1967), page 149.

3.14159 26535 89793 23846 26433 83279 5

May I have a month, professor,
To figure these, you brain assessor?
Calculate, student, calculate now!
As the figuring gets longer,
My friend, hope you get stronger
And no figures, incorrect, allow.

—Aaron L. Buchman, *School Science and Mathematics,* vol. 53 (February 1953), page 106.

3.14159 26535 89793 23846 26433 83279 5

You I sing, O ratio undefined
By strict assay and lined,
Sequence limitless. Stunned regarding you,
We see eternity—alas—unwind
In random cast and rue,
Dejected out of measure, reckoning blind.

—John Freund (English Department, Grand Valley State College, Allendale, Michigan), *The Mathematics Teacher,* vol. 62 (April 1969), page 348.

3.14159 26535 89793 23846 26433 83279 5

May I tell a story purposing to render clear
the ratio circular perimeter-breaths, revealing
one of the problems most famous in modern days,
and the greatest man of science anciently known.

—C. J. Jackson, *Mathematical Gazette,* vol. 4 (July 1907), page 103.

One of the longest mnemonics ever written for pi was published in the *Journal of Recreational Mathematics* (vol. 8, no. 3, 1975–1976, page 226) by Michael Keith of Hazlet, New Jersey. The mnemonic is a 216-word story of a fictional exploration of Mars, entitled "To Explore A Memorial to Martians." It begins as follows:

For a time I tried exploring in gloomy shade. The
thick darkness descended quickly. Tenseness lay
in the twilight.

The story ends with "The End," as the 215th and 216th digits of pi are both 3. The author used the convention of letting every punctuation mark (other than a period) stand for zero. In certain places, single words of 10, 11 or 12 letters are used to express two digits: 1, 0 or 1, 1 or 1, 2. Curiously, the title is a six-digit mnemonic for *e*.

As impressive as Keith's 216-digit mnemonic is, a longer pi mnemonic for 402 decimal places was published in the *Mathematical Intelligencer* (vol. 8, no. 3, 1986). The mnemonic began: "For a time I stood pondering on circle sizes." So who is the author of this new record holder? Why, none other than Michael Keith himself. The editors remarked, "We have seen pi-mnemonic sentences, poems, and now, a short story. Perhaps someday a complete novel?" Well, perhaps not. But if you are tempted to create a longer mnemonic, you should know that at decimal place 601, the first triple zero occurs, and at decimal place 772, the sequence 9999998 occurs. This seven-digit sequence contains the largest seven-digit sum among the first million decimals. Pi worked out to 10,000 digits can be found as an appendix to Petr Beckmann's delightful *A History of Pi* (St. Martin's Press, 1971).

$1/\pi \ (= 0.3183098\ldots) \approx 0.318310$

Can I discover the reciprocal?

> —John Sturgeon Mackay, *Proceedings of the Edinburgh Mathematical Society,* vol. III (May 1885), page 106.

$1/\pi = 0.3183098$

O! how I remember you: O
difficult equation.

> —E. Leedham (Buenos Aires), *Mathematical Gazette,* vol. 10 (October 1921), page 323.

$\log \pi = 0.49714\ 98726\ 94134$

This Logarithm employs a zero
character; mantissa follows in digits
precisely what I now give.

> —Alan S. Hawksworth, *American Mathematical Monthly,* vol. 38 (March 1931), page 158.

Algebraic Mnemonics or Please Pardon My Dear Aunt Sally

The curious phrase in the title is an old mnemonic for the order of the algebraic operations. The initial letter in each word is identical to the first letter in the following: Parentheses, Powers, Multiplication, Division, Addition, and Subtraction. Two variations are: *Please Excuse My Dear Aunt Sally* (E = exponents), and *Bachelors Please My Dear Aunt Sally* (B = brackets).[2]

---

[2] Note that implied multiplication is done before division and before explicit multiplication: $1 = 4/2n$, where $n = 2$, but $4 = 4/2 * 2$.

An old device for teaching binomial multiplication is FOIL = **First, Outside, Inside, Last**: $(a+b)(c+d) = ac + ad + bc + bd$.

Which is correct: $x^{3/4} = \sqrt[4]{x^3}$ or $x^{3/4} = \sqrt[3]{x^4}$? The first one is correct, of course. Here is how to remember it: The 3 is on top and in *power*; the 4 is underneath like a tree *root*. —Samuel H. Barkan, *Mathematics Teacher* (February 1934), pages 94–95.

Humor can sometimes be used with effect as a mnemonic. When students showed difficulty with the words *numerator* and *denominator,* Barkan told them the following: "The word *denominator* begins with the letter *d* as in the word *down,* and the word *numerator* begins with the letter *n* as in the word *nup.*"

Question: Name this tune! (See Figure 13.)

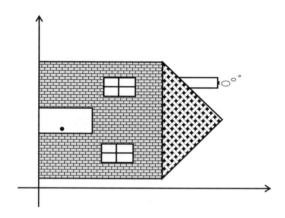

FIGURE 13

Answer: "Home on the range." This pun is silly and childish, but it will help some students remember that, in a high school algebra class, the x-axis is the domain, and the y-axis is the range. I have used most of these algebraic mnemonics in my classroom.

The British poet W. H. Auden (1907–1973) won first prize for mathematics at St. Edmund's School in Hindhead in Surrey, England when he was 13. In later years he regretted that he did not continue his study of the subject. He recalled being asked to learn the following mnemonic around 1919:

Minus times Minus equals Plus;

The reason for this we need not discuss.[3]

---

[3] Humphrey Carpenter, *W. H. Auden, A Biography* (Houghton Mifflin, 1981), pages 17 and 23.

Beginning algebra students often wonder why minus times minus is plus. I think the reason for this is very much worth discussing. The "rule of signs" ($-1 \times -1 = 1$), together with all other definitions governing integers and fractions, are postulates that cause the results of mathematical calculations to correspond with physical reality. For example, had we ruled that $(-1) \times (-1) = -1$, then $-1(1 + -1) = -1 + -1 = -2$. But to avoid contradiction with earlier definitions—that were chosen to correspond with reality—we need $-1(1 + -1) = -1(0) = 0$. In this way, the rule of signs extends the distributive law from positive integers to all integers.[4]

The Quadratic Formula:

$$x = \frac{-b \pm \sqrt{b^2 - 4ac}}{2a}$$

Since "quad" means "four," students often wonder why mathematicians chose the word "quadratic" to describe an equation with an exponent of two. The "quad" in quadratic refers to the x^2 term which is a square. Incidentally, there are other forms of the quadratic formula. Given the quadratic equation $ax^2 + bx + c = 0$, we may conclude the new quadratic formula:

$$x = \frac{-2c}{b \pm \sqrt{b^2 - 4ac}}.$$

Rationalize the denominator, and you get the "old" quadratic formula. This example helps to explain the old rule: "Always rationalize your denominators." Why? Because it is difficult to tell if two expressions are equal when they are in different forms. That is also the reason we insist on reducing fractions.

> From square of b take $4ac$;
> Square root extract, and b subtract;
> Divide by $2a$; you've x, hooray!
> > —Author unknown, *Proceedings of the*
> > *Edinburgh Mathematical Society,* vol. III
> > (1885), pages 106–107.

---

4 This point is discussed in detail by Richard Courant and Herbert Robbins in *What Is Mathematics?* (Oxford, 1941), pages 54–56.

When you have written $-b$,
The double sign put down;
Then $b^2 - 4ac$.
With square root mark your crown;
Beneath it all a line you trace,
Beneath which line $2a$ you place.

> —N. D. Beatson Bell, *Proceedings of the Edinburgh Mathematical Society,* vol. III (1885), page 107.

Trigonometry

Lord God of Hosts, be with us yet,
Lest we forget—lest we forget!

> —Rudyard Kipling, *Recessional*

Lord God of Hosts was with me not,
For I forgot—for I forgot.

> —Anonymous

I always show my geometry students the old trig mnemonic "Some Of Her Children Are Having Trouble Over Algebra" ($\sin A = $ opp/hyp; $\cos A = $ adj/hyp; $\tan A = $ opp/adj). Then I ask them to construct their own mnemonics. The next day, I collect the mnemonics and read the interesting ones to the class. The following are my favorites.

SOHCAHTOA!

1. So often have cruel and hateful tricks opened arguments.
   > —Sam Feikema, 1991
2. Some out-houses can actually have totally odorless aromas.
   > —Michael Gibson, 1989
3. She offered her cat a heaping teaspoon of acid. —Allison Morgan, 1991
4. Shoplifting often happens. Cathy admitted her theft one afternoon.
   > —Guilia Perazzoli, 1980
5. Stueben owes his class a hardy ton of A's. —Stephanie Deller, 1980
6. Stueben's original homework causes awful headaches to over-achievers.
   > —Becky Beasley, 1989
7. Soaring over Haiti, courageous Amelia hit the ocean and . . .
   > —Lisa Wakeham, 1980
8. Stamp out homework carefully, as having teachers omit assignments.
   > —Stephanie Deller, 1980.

9. School! Oh how can anyone have trouble over academics.

—David McCulloch, 1980.

10. Some of her children are having trouble over algebra.

—Anonymous, pre-1968

11. Smart Children Think: Oscar had a heap of apples. Or (to the tune of Yankee Doodle Dandy) Oscar had a heap of apples, sine and cosine tangent.   —Anonymous

12. Sing out happily, 'cuz a healthy tune obtains applause.

—Andrea Agostini, 1989

13. Students often hardly concentrate after hearing the only answer.

—Andrea Agostini, 1989

14. Sin—oh!
    Cos—ah!
    Tan—oh/ah!   —T. Percy Nunn, 1928

15. Oh heck! Another hour of algebra *with* some crazy teacher.

—Dictionary of Mnemonics, 1972

16. Sine-off helps congestion and headaches. Take one anyway.

—J.B. Wilson, 1980

In which quadrants do the six trig functions take positive values? Figure 14 provides the answer, but how shall we remember it?

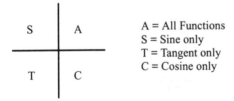

FIGURE 14

Starting in the first quadrant and proceeding counterclockwise (like an angle going from 0° to 360°), we say: *All Students Take Calculus* or *Awful Stinky Trig Course* or *Are Simpletons Teaching Courses?* or *Are Schools Test-Crazy?* or *All Schools Torture Children.* Since a function and its reciprocal have the same sign, the letters ASTC (or the ACTS mnemonic or the iron CAST rule) also indicate the signs in the different quadrants of the cosecant, secant, and cotangent.

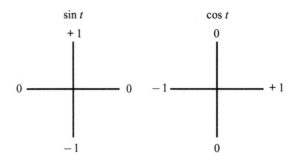

FIGURE 15

Notice that the middle letter of sin is i (looks like 1) and the middle letter of cos is o (looks like zero). If we place these words above the positive x-axis, then these diagrams become easier to memorize (see Figure 15).

Another curious mnemonic is this chart:

θ	$0°$	$30°$	$45°$	$60°$	$90°$
$\sin\theta$	$\sqrt{\dfrac{0}{4}}$	$\sqrt{\dfrac{1}{4}}$	$\sqrt{\dfrac{2}{4}}$	$\sqrt{\dfrac{3}{4}}$	$\sqrt{\dfrac{4}{4}}$
$\cos\theta$	$\sqrt{\dfrac{4}{4}}$	$\sqrt{\dfrac{3}{4}}$	$\sqrt{\dfrac{2}{4}}$	$\sqrt{\dfrac{1}{4}}$	$\sqrt{\dfrac{0}{4}}$

What are the radian values of the standard angles $120°$, $135°$, and $150°$? Answer: Take the middle digit, divide by the following consecutive digit, and append π: $2\pi/3$, $3\pi/4$, $5\pi/6$. (Thanks to C. B. Brown, my former precalculus student.)

Radians and Degrees

[E]ven after weights came into use it was not the custom to speak of such a fraction as 3/4 of a pound. The world avoided difficulties of this kind by creating such smaller units as the ounce and then speaking of the particular number of ounces. ... In fact, the origin of such compound numbers as 3 yd. 2 ft. 8 in. is to be sought in the effort of the world to avoid the use of fractions. —David Eugene Smith, *History of Mathematics,* vol. 2 (1925; Dover reprint, 1958), pages 208–209.

Students often ask why the number 360 was chosen to be the number of degrees in a circle. The best answers I could find are these: a) the number 360 has many convenient factors: 2, 3, 4, 5, 6, 8, 9, 10, 12, 15, 18, 20, 24, 30, 36, 40, 45, 60, 72, 90, 120, 180; b) ancient Babylonian mathematics used a base 60 system; and c) angle measurement was needed in astrology and astronomy, and the year has almost exactly 360 days. According to the *Oxford English Dictionary,* the word radian did not appear in print until 1873.

> The reason (and the only reason) for measuring angles in radians rather than degrees is precisely the same as the reason for preferring logarithms to base e to common logarithms—namely, that theoretical arguments become much simpler. For example, the "differential coefficient" of $\sin \theta$ is $\cos \theta$ if θ is the number of radians in an angle while it is $\dfrac{\pi}{180}\cos \theta$ if θ is a number of degrees. —T. Percy Nunn, *The Teaching of Algebra* (Longmans, Green, 1923), page 499.

Given a circle of radius r and central angle θ, the length of the intercepted arc is $r\theta$ in radians and $\dfrac{180r\theta}{\pi}$ in degrees. This is another illustration of the fact that radian measure simplifies trigonometric expressions and formulas. To convert degrees into radians or vice-versa, we use $\dfrac{\pi}{180}$ and $\dfrac{180}{\pi}$ as conversion factors. But how does a student remember which one to use? To convert degrees to radians, cancel out the degree measure:

$$90° = 90° \times \frac{\pi}{180°} = \frac{\pi}{2} \text{ radians.}$$

To convert radians to degrees, cancel out the radian measure (by dividing by π radians):

$$\frac{\pi}{2} \text{ radians} = \frac{\pi}{2} \times \frac{180°}{\pi} = 90°.$$

Alison's Diagram

If one writes the names of the six trigonometric functions in a particular fashion, as in Figure 16, an amazing thing happens: All possible true, trigonometric identities of the form a) $x \div y = z$ or b) $x \times y = z$ are represented, where x, y, z are the basic trigonometric functions of the same angle t. This is how it works: Observe the six trigonometric names on the perimeter of the star. Examine any three consecutive functions—e.g., $\tan t$, $\sin t$, and $\cos t$. Then the product of the ends always equals the middle—i.e., $\tan t \times \cos t = \sin t$. Also, the middle function divided by one of the end functions is equal to the

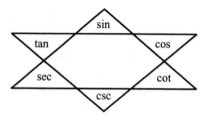

FIGURE 16

other end function—i.e., $\sin t \div \tan t = \cos t$, and $\sin t \div \cos t = \tan t$. Thus, at a glance, one can read off a list of 18 simple relations [1].

Years ago I discovered that a bit of positive social commentary in a lesson quickly attracts the student's attention and sometimes leads to worthwhile discussions. Here is my effort to introduce the trigonometry sum and difference formulas in an interesting way:

As we all know, some of the people to whom we are attracted are not attracted to us. And it is not unusual for a person who has shown interest in us to later lose interest in us. Maybe that is a good thing, because it forces us to date a lot of people and to become more experienced in maintaining relationships. Anyway, this is the story of Sinbad and Cosette. Sinbad loved Cosette, but Cosette did not feel the same way about Sinbad. Naturally, when Sinbad was in charge of their double date, he put himself with Cosette, and he put his brother with her sister:

$$\sin(A + B) = \sin A \cos B + \cos A \sin B.$$

$$\sin(A - B) = \sin A \cos B - \cos A \sin B.$$

Sinbad loved to tell people that his and Cosette's signs were the same. However, when Cosette was in charge of the double date she placed herself with her sister and put Sinbad with his brother. She made sure everyone knew that their signs were NOT the same:

$$\cos(A + B) = \cos A \cos B - \sin A \sin B.$$

$$\cos(A - B) = \cos A \cos B + \sin A \sin B.$$

Also, notice that Cosette placed herself and her sister BEFORE Sinbad and his brother. This detail was important to Cosette. She was very snobby, you know.

The point of this mnemonic is the fun, and then, maybe, some help to the memory.

"Lemniscate . . . looks like eight."
>—A mnemonic rhyme by my precalculus
>student Uma Rao (1996).

Geometry

For circumference and area:

Fiddle de dum, fiddle de dee,
A ring round the moon is π times d;
But if a hole you want repaired,
You use the formula πr^2.
>—From L. A. Graham, *Ingenious Mathematical
>Problems and Methods* (Dover, 1959), page 11.

$$e = 2.7\ 1828\ 1828\ 45\ 90\ 45\ 2353602874\ldots$$

The symbol for e was first used by Leonhard Euler (c. 1728) who may have selected it to stand for "exponential."[5] Below, the number of letters in each word represents a corresponding digit of e.

2.7 1828 1828
It enables a numskull to memorize a quantity of numerals.
>—Gene Widhoff.

2.7 1828 18284
To express e, remember to memorize a sentence to simplify this.
>—John L. Greene.

Both of the preceding mnemonics appeared in Martin Gardner's *Mathematical Puzzles and Diversions* (Simon and Schuster, 1959), page 109.

2.7 1828 1828
I'm forming a mnemonic to remember a function in analysis.
>—Maxey Brooke, in *The Unexpected Hanging,* by Martin
>Gardner (Simon and Schuster, 1969), page 40.

---

5 See David Eugene Smith, *A Source Book in Mathematics* (1929; Dover reprint 1959), pages 95–98 which contains Cajori's translation of several of Euler's papers that use e. Unfortunately, Euler never does give even a hint as to why he chose e.

2.7 1828 1828

It repeats: A constant of calculus,

A constant of calculus.
>—Jeffrey Strehlow, *Omni,* vol. 2 (August
>1980), page 122.

2.7 1828 1828 459

To express *e* properly is possibly a proposal
to consider logs based naturally.
>—Don K. Poulson, *Omni,* vol. 2 (August
>1980), page 122.

2.7 1828 18284 59

We proffer a mnemonic to remember a
standard or Naperian base value instantly.
>—Dr. John Sturgeon MacKay, *Proceedings of
>the Edinburgh Mathematical Society,* vol. 3
>(1885), page 106.

The author of the following item used the word *nothing* to stand for a zero and used the twelve-letter phrase *what's-his-name* to represent the digits 1 and 2.

$\mu = \log e = 0.43429\ 44819\ 03251\ 82765\ 11289\ 18$

Base ten: Best in practical work. Can't evaluate a logarithm?
Nothing can be nicer! A constant is clearly needed first.
A what's-his-name (Naperian) logarithm I evaluate.
>—William Hope-Jones, *The Mathematical
>Gazette,* vol. 10 (October 1921), page 328.

Various Numbers

Euler's Constant $\gamma = 0.57721\ 56649\ldots$

These numbers proceed to a limit
Euler's subtle mind discerned.
>—Morgan Ward, *American Mathematical
>Monthly,* vol. 38 (November 1931), page 522.

The golden mean is sometimes represented by the Greek symbol tau (τ) in British publications and by the Greek phi (ϕ) pronounced either fee or fie in American publications. This symbol (ϕ), with its vertical or near-vertical stroke, is not to be confused with the symbol for the null set, the Danish and Norwegian ∅, which has a 45° stroke. The symbol for the null set was introduced by André Weil in a Bourbaki publication (c. 1936).[6]

$\phi = 1.618\ldots$
A (decimal) number I remember.
                              —E. Kim Nebeuts (1980).

$1.414 = \sqrt{2}$ I wish I knew
            (the root of two).

$1.732 = \sqrt{3}$ O charmed was he
            (To know root three).

$2.236 = \sqrt{5}$ So we now strive
            (To find root five).
                              —*Dictionary of Mnemonics* (London, 1972).

Miscellaneous Mnemonics

Here is a useful mnemonic for remembering the relative order of the metric prefixes: *Kids have dropped over dead converting metrics.*

> *kilo* (10^3)
> *hecto* (10^2)
> *deka* (10^1)
> *original* meters, liters, grams, etc. $(10^0 = 1)$
> *deci* (10^{-1})
> *centi* (10^{-2})
> *milli* (10^{-3})

It is easy to distinguish between *deka* (10^1) and *deci* (10^{-1}), because the *k*-sound is also found in *kilo* (> 1), and the *c*- and the *i*-sounds are found in *centi* (< 1).

In 1979, one of my pre-calculus students carried the name Edwin Malcolm Clayton C——. His father was a physicist, he explained to me one

---

6 André Weil, *The Apprenticeship of a Mathematician* (Birkhäuser, 1991), page 114.

day, and had constructed the name in honor of Einstein's famous equation $E = MC^2$.

The antique Josephus Problem was studied by Euler and other famous mathematicians:

> A ship containing 15 treacherous crew members and 15 cooperative crew members was caught at sea in a furious storm. In order to lighten the ship, the captain decided that half of the crew was to be thrown overboard. To this end, the crew was placed in a circle, and starting the count from a specific crew member every ninth man was sacrificed to the sea. The counting continued around the circle until half of the crew was sacrificed. To be saved, where should the cooperative crew members have placed themselves?

The answer is given in Figure 17, where C stands for a cooperative crew member, and T stands for a treacherous crew member.

```
Position of 1st man
          ↓
Answer:   CCCC TTTTT CC T CCC T C TT CC TTT C TT CC T
```

FIGURE 17

One of the most lyrical mnemonics ever written was devised to help remember the ordering of the crew (4C, 5T, 2C, 1T, etc.) so that only the treacherous crew members were thrown overboard. In the rhyme below, each vowel gives us a number in the solution ($a = 1$, $e = 2$, $i = 3$, $o = 4$, $u = 5$).

> From Number's Aid And Art,   (= 4, 5, 2, 1, 3, 1, 1)
> Never Will Fame Depart.   (= 2, 2, 3, 1, 2, 2, 1)
> > —Rev. Edward Samuel Taylor, *The History of Playing Cards* (London, 1865, reprinted by Charles E. Tuttle), page 485.

The following is a mnemonic my father taught me in 1954: "When multiplying by 9, the first digit of the answer is one less than the digit being multiplied by 9. Then just add enough to make your answer's digits sum to 9." I still use it! Years later I came across the same rule in a book published in London in 1893.

The Poisson distribution states that if m happenings occur on the average, then the probability that n happenings will occur is given by: $P_m(n) = m^n e^{-m}/n!$. Or to say this another way, "m to the n, e to the $-m$, over $n!$ is correct." or the word *mnemonic*, itself! —Manuel H. Greenblatt, *Mathematical Entertainments*, (Thomas Y. Crowell, 1965), pages 122–123.

Finally, we have a mnemonic for remembering how to spell *mnemonics:*

Mnemonics neatly eliminate man's only
nemesis: insufficient cerebral storage.

—William D. Harvey, *Omni*, vol. 2 (August
1980), page 122.

With reason sound the Greeks define,
As mother of the Muses nine,
Mnemosyne—an *alias* grand
For simple Memory to stand.
No one without her could learn aught.
No one without her could be taught.
She grants her graces without dearth
To those who estimate her worth.

—James Copner, *Memoranda Mnemonica* (London:
Williams and Norgate, 1893), pages 393–395.

Notes

1. This is known as "Alison's Diagram" after its publication by John Alison in 1886 in the *Proceedings of the Edinburgh Mathematical Society,* vol. IV (June 11, 1886), page 88. It has been rediscovered and republished many times since: See L. H. Rice, *The Mathematics Teacher,* vol. 5 (March 1913), pages 145–146, and Pincus Schub, *Scripta Mathematica,* vol. 20 (September-December, 1954), page 217. George Wentworth and D. E. Smith recommended memorizing Alison's Diagram in their textbooks, which were popular from before 1900 to about 1940. I used the diagram as a freshman in college in 1968 to help solve trigonometric identities.

9
Academic Humor

A mathematician died and went to Heaven. He was given a modern office with a wonderful view and a faultless secretary. Almost immediately he began to turn out brilliant papers. After six months he realized that none of his papers had reached print in the several heavenly publications of mathematics. So he asked St. Peter how long the refereeing process would take.

"Oh, about six years," said St. Peter.

"Six years! That's almost intolerable. If it takes six years in Heaven, then how long does it take in Hell?"

"I don't know," said St. Peter. "Why not go down there and find out?"

So the mathematician took a tour of Hell. Eventually he met his hellish counterpart and asked how long it took to get a paper published.

"Well, if you use e-mail—and who doesn't—it can be done in two days."

"Two days. That's amazing! In Heaven it takes about six years. I don't understand how you guys can get things done so fast."

"Well," said his counterpart, "maybe it's because most of the editors and referees are down here."

Before you walk, you must crawl. And before you master complex analysis, you must be on Ahlfors. —Michael H. Brill, an industrial physicist, in a conversation in 1992 [1].

Mike Brill once mentioned that he had never been motivated to study chess because there was so much to memorize. I suggested that his reason wasn't valid and said: "You studied physics, and there is much to memorize there. For example, you must have memorized Planck's constant." After a moment's pause, he said: "Yes, it's h, isn't it?" Incidentally, David Mehler, chess teacher and director of the U.S. Chess Center in Washington, D.C., told me that he thought memorizing openings actually made a chess player weaker.

The index of Thomas' *Calculus and Analytic Geometry* (4th ed.) contains a reference to whales. Upon turning to the indicated page, one finds nothing written about whales. Instead, there are two graphs whose shapes coincidentally resemble whales.

I am now going to tell you what I think is the best academic pun ever made. But it is so academic that it requires some background. First, *Deism* is a form of theology that was popular among intellectuals in the late 18th and early 19th centuries. Second, part of the legacy of the Newton-Leibniz controversy was the insistence of the English on using the inferior fluxions (*dot-notation*) of Newton for the differentials (*d-notation*) of Leibniz. Now here is the pun:

> The rapid rise in French and German mathematical research had made little impression in Great Britain. Faced by this situation, a group of young Cambridge mathematicians in 1812 formed what they called the Analytical Society. In the words of Charles Babbage (1792–1871), one of the leaders, the aim of the society was to promote "the principles of pure d-ism as opposed to the dot-age of the university." —Carl B. Boyer, *A History of Mathematics* (John Wiley, 1968), page 583.

The Royal Chain Mail Factory had received a large order for battle uniforms. Each uniform consisted of a toga and a pair of short pants. The tailor's only problem was how long to make the pants: too short, and a soldier could be exposed; too long, and a uniform would be excessively heavy. So they called in a mathematician. He had a uniform made and tested. The hem on the pants proved to be too short, so he increased it a little bit, then a little more, and then a little bit more, and so on until finally he was able to derive an exact trousers-length depending on the leg-length of the soldier. The chief tailor was curious. "How did you determine this ratio?" he asked.

"Easy, " said the mathematician. "I just used the "Wire-Trousers Hem Test of Uniform Convergence." This pun on the Weierstrass M-test of uniform convergence was sent to me by Andrius Tamulis.

If students do not understand a joke immediately, they may not appreciate it. That is why some of the more academic puns should be told only in a class that has recently been working on material referenced by the joke. Why did Weierstrass choose M? The sci.math usenet poster Don Heller suggested the answer may be in the national origin of the writer: S = set (English), E = ensemble (French), M = menge (German).

Dr. Carl E. Linderholm once wrote a humorous book which looked at elementary mathematics from the point of view of advanced mathematics: *Mathematics Made Difficult* (London: Wolfe Publishing, Ltd., 1971). Here are a few of the gems:

1. A diagram is worth a thousand proofs.
2. Half pi should be called *hi* and be written τ.
3. Mathematicians pretend to count by means of a system supposed to satisfy the so-called Peano axioms. In fact, the piano has only 88 keys; hence, anyone counting with these axioms is soon played out.

In a circle, the ratio of the circumference to the diameter is constant and is called "pi." In an ellipse, this ratio is not constant, and is related to the eccentricity of the ellipse. In the 18th century the French Geometer Pierre Meringue wrote a book on ellipses and discovered a useful theorem that enabled him to deduce over 149 propositions concerning the eccentricity of ellipses. Today in France the eccentricity of the ellipse is called Meringue pi, and his theorem is often referred to as the "Lemma on Meringue pi." —A variation of these puns appeared in *Mathematics Magazine,* vol. 66, no. 5 (December 1993), page 321.

Antoni Zygmund once asked if the *World Series* shouldn't be called the *World Sequence?* And shouldn't a *combination* lock be called a *permutation* lock? John Von Neumann once had a dog called *Inverse.* It would sit when told to stand and go when it was told to come. Von Neumann pronounced the term *infinite series* as *infinite serious.*

Noitaton Hsilop esrever diova! —E. Kim Nebeuts.

I once wrote to Martin Gardner to ask what the strange coded dedication meant in his book *Martin Gardner's New Mathematical Diversions From Scientific American* (Simon and Schuster, 1966). The answer he sent back was so simple that I was embarrassed not to have decoded it myself. Perhaps the reader can do better than I did. Here is his dedication:

> Evoly met
> TO L. R. AHCROF,
> emitero meno.

Alberto Torchinsky of the University of Indiana has collected some of the statements he has heard lecturers make over the years:

1. As you can see, these equations are very easy to remember ... Hold on ... I've missed a term.
2. I worked it out last night. Let's see if it works in the daylight.
3. Watch this proof carefully. It looks like a confidence trick.
4. This will enable you to pull derivatives through integrals, which you have wanted to do all your life—and some of you have been. This tells you when you can do it legally.

Joel Chan, former editor of *MAT 007 I News*, the University of Toronto's "wreckreational math news letter (published bioccasionally)" recorded the following remarks from Canadian mathematical lectures.

1. These handouts I give you, they are not souvenirs.
2. You can check this in the privacy of your own home.
3. Sorry, I used gamma instead of alpha ... no, gamma instead of lambda ... no, ...
4. Would you like a take home? You don't know what you are getting into.
5. Assume $\mathbb{R}$ is a principal ideal domain, because life is too tough otherwise.
6. Sometimes dirty proofs are better because one understands them better.
7. If you have this notation, you don't have to think.
8. I'm going to give you a definition. Then I'm going to tell you why it's wrong.

The following items are found in Donald Knuth, Ronald Graham, and Oren Patashnik, *Concrete Mathematics,* 2nd ed. (Addison Wesley, 1989).[1]

---

[1] *Concrete* mathematics is a blend of *continuous* and dis*crete* mathematics.

1. Where the pessimist sees a half-closed interval, the optimist sees a half-open interval.
2. One of the occupational hazards of teaching calculus is d generating function dz's.[2]
3. Question: What does a drowning analytic number theorist say?
   Answer: Log-log, log-log, ...
4. Question: Can you prove Lagrange's identity?
   Answer: No. If it is a true identity, it doesn't need to be proved. If it's a case of mistaken identity, then it can't be proved. And besides, it's especially difficult to prove the identity of a person who has been dead for nearly 200 years.
5. Question: The sigma sign occurs more than 1000 times in Knuth's book *Concrete Mathematics*. Can you name another book in which it appears so often?
   Answer: The *Iliad*.

Misunderstandings in the History of Mathematics

In the interest of historical accuracy let it be known that ...
1. Michael Rolle was *not* Danish, and did *not* call his daughter "Tootsie."
2. William Horner was *not* called "Little-Jack" by his friends.
3. The "G" in G. Peano does *not* stand for "grand," and Peano arithmetic is not done modulo 88.
4. René Descartes' middle name is *not* "push."
5. Isaac Barrow's middle name is *not* "wheel."
6. Fibonacci's daughter was *not* named "Bunny."[3]
7. There is no such place as the University of Wis-cosine, and if there were, the motto of their mathematics department would *not* be "Secant ye shall find."
8. Although Euler is pronounced oil-er, it does *not* follow that Euclid is pronounced oi-clid. [After posting this on the Internet, I was informed that indeed this is exactly how Euclid is pronounced in German and some other languages. Oops!]

---

[2] Degenerating function disease. Get it?

[3] Leonardo Fibonacci's original problem in 1202 that used the Fibonacci sequence was a problem about breeding bunnies.

9. Franklin D. Roosevelt *never* said, "The only thing we have to sphere is sphere itself."

10. Fibonacci is not a shortened form of the Italian name that is actually spelled: F i bb ooo nnnnn aaaaaaaa cccccccccccccc cccccccccccccccccccc-ccc iiiiiiiiiiiiiiiiiiiiiiiiiiiiiiiiiiii

11. It is true that August Möbius was a difficult and opinionated man. But he was *not* so rigid that he could only see one side to every question. —Sidney Harris, *More Cartoons From Sidney Harris* (Freeman, 1991), no page number.

12. It is true that Johannes Kepler had an uphill struggle in explaining his theory of elliptical orbits to the other astronomers of his time. And it is also true that his first attempt was a failure. But it is not true that after his lecture the first three questions he was asked were "What is elliptical?" What is an orbit?" and "What is a planet?" —Sidney Harris, ibid.

13. It is true that primitive societies use rough approximations for the known constants of mathematics. For example, the northern tribes of Alaska consider the ratio of the circumference to the diameter of a circle to be 3. But it is *not* true that the value of 3 is called Eskimo pi.

14. It is true that Sir Isaac Newton was born and raised in a farm district. But it is *not* true that he chose the term COW-culus because the mathematics described the paths of MOO-ving objects.

Titles of Papers

The forthcoming issue of *Trivia Mathematica* will include the following papers:

1. Certain invariant characterizations of the empty set.
2. Proof that every polynomial inequality over the complex numbers has at least one unroot.
3. On unprintable propositions.
4. The impossibility of a proof of the impossibility of a proof.
5. $P = NP$, where $N = 1$.
6. Groups of order 1.
7. The Null Graph and Other Pointless Concepts.
8. What is an Answer?

Items 1–4 are by Norbert Wiener and Aurel Wintner (1939), in Peter Duren, *A Century of Mathematics in America,* vol. 1 (American Mathematical Society, 1988), page 95. The other items were obtained from the sci/math newsgroup on the Internet.

A Mathematical Q&A

Q: What is the world's longest song?

A: Aleph Naught[4] Bottles of Beer on the Wall: "Aleph-naught bottles of beer on the wall, Aleph-naught bottles of beer. Take one down, pass it around, Aleph-naught bottles of beer on the wall. Aleph-naught bottles of beer on the wall, ..."

Q: What has white fur, pink ears, hops around, and relates the curvature of a closed two-surface manifold to its topology?

A: The Gauss-Bunny theorem. [This is a pun on the Gauss-Bonnet theorem.]

Q: What swims in the ocean and has integer solutions?

A: A Dolphin equation. [A Diophantine (dy·uh·FAN·tyn) equation is an indeterminate equation with integer coefficients and integer solutions.]

Q: What is big, gray, and has integer solutions?

A: An Elephantine equation. —Norbert Wiener and Aurel Wintner (1939).

Q: What is tall, has a long spotted neck and is practiced in freshman algebra?

A: Giraffical representation. —Norbert Wiener and Aurel Wintner (1939).

Q: Why did the chicken cross the road?

A: The classic answer to this children's question is well known, "To get to the other side." When I asked my wife this question, she replied, "To get away from Frank Perdue." (Frank Perdue is an East-coast chicken rancher.) In the August 1994 issue of *Word Ways*, Susan Thorpe noted that the phrase "the chicken crosses the road" anagrams into "she checks corn at other side."

Physicist's answer: It was attracted to a chicken of opposite gender on the other side of the road, or it was repelled by a chicken of same gender on the same side of the road.

Mathematician's answer 1. (number theory): According to the chicken-hole principle it had to be on other side. 2. (analysis): Because the chicken wanted a vector perpendicular to the road. What do you get when you cross a chicken with a banana? Answer: Chicken banana sine theta in a direction mutually perpendicular to the two as determined by the right-hand rule. This is the vector cross-product (= vector product = outer product) rule.

The Bad Lecturer's answer: If you saw me coming, you would cross the road, too.

---

[4] Aleph-naught, aleph-null, or aleph-zero ($\aleph_0$) is the symbol representing the cardinal number of a countably infinite set (e.g., the positive integers, the rational numbers, etc.) and is also called the power of a set. The cardinal number of the set of real numbers (designated c) is greater than aleph-null. And the cardinal number of the set of all subsets of real numbers (designated 2^c) is greater than c.

Topologist's chicken question: Why did the chicken cross the Möbius strip? Answer: To get to the other ... uh ... uh ... (pause) never mind.

Q: What do you get when you cross a vector with a mountain climber?
A: Nothing, because the mountain climber is a scalar [2].

Q: What is yellow and equivalent to the axiom of choice?
A: Zorn's lemon [3].

Q: Why didn't Newton prove the non-existence of an algebraic solution to the general quintic equation?
A: He wasn't Abel.

Q: What is purple and commutes?
A: An Abelian grape. [Note: There are two pronunciations of the term Abelian: uh·BEEL·yuhn and uh·BEE·lee·uhn. The last name of Neils Henrik Abel (1802–1829) is pronounced AHB·uhl not A·bull. An excellent source for correct pronunciations is Merriam-Webster's *Biographical Dictionary*.]

Q: What is purple, commutes, and has only a limited number of worshipers?
A: A finitely venerated Abelian grape.

Q: What is purple, commutes, and gets great gas mileage?
A: A compact Abelian grape.

Q: What is purple, commutes, and is environmentally conscious?
A: A cyclic Abelian grape.

Q: How do group theorists show appreciation to other mathematicians?
A: By saying, "Thanks. Thanks Abelian."

Q: What is yellow and expressible in a power series?
A: A bananalytic function [4].

Q: What is yellow, normed, and complete?
A: A bananach space.

Q: What is large, gray, and probably has a real part equal to 1/2?
A: The Riemann Hippopotamus, not to be confused with the Continuum Hippopotamus.

The Professor and the Student

Professor: What is the greatest cardinal?
Student: The Pope! (chuckle, chuckle)
Professor: I don't think so.
Student: Why not?
Professor: Every Pope has a successor.

Student: I know a trick way to tell if an integer is evenly divisible by a 7.
Professor: So do I. Write it in base 7, and then look at the last digit.

Student: (After deriving $F = -ma$). Oops! Well, I've only made one error with a sign.
Professor: Or an odd number of them.[5]

Symbol Play

$$\text{Happiness} \in \mathbb{R} = \int_{\text{service}}^{\text{others}} \text{accomplishment } d(\text{opportunity}) = \frac{d(\text{dedication})}{d(\text{time})}.$$

Translation: Real happiness is an integral function of accomplishment from service to others with respect to opportunity and is a derivative of dedication over time.

Lewis Carroll (an Oxford professor of mathematics) published the following stanza in his book of poetry: *Phantasmagoria and Other Poems* (1869).

> And what are all these mysteries to me
> Whose life is full of indices and surds?
> $x^2 + 7x + 53 = 11/3.$

$$\int_1^{\sqrt[3]{3}} z^2 dz \cos \frac{3\pi}{9} = \ln \sqrt[3]{e}$$

Integral z squared dz
from one to the cube root of three
    Times the cosine
    Of three π over nine
equals log of the cube root of e.
         —Betsy Devine and Joel E. Cohen, *Absolute Zero Gravity* (Simon & Schuster, 1992), page 37.

---

[5] Thanks to Ba Thinh Van, my former computer science student.

$$\int_0^{\pi/6} \sec y \, dy = \ln \sqrt{3}$$

The integral $\sec y \, dy$
From zero to one-sixth of pi
Is the log to base e
Of the square root of three.
Um ... times the square root of the fourth power of i.
—Anonymous

Leigh Mercer's Symbol Rhymes

1, 264, 853, 971 . 2758463
or
One thousand two hundred and sixty
four million eight hundred and fifty
Three thousand nine hun-
Dred and seventy one
Point two seven five eight four six three.
—Leigh Mercer, in Martin Gardner, *The
Unexpected Hanging* (Simon and Schuster,
1969), page 238.

$$\frac{12 + 144 + 20 + 3\sqrt{4}}{7} + 5(11) = 9^2 + 0.$$

or
A dozen, a gross, and a score
Plus three times the square root of four
Divided by seven
Plus five times eleven
Is nine squared and not a bit more.
—Leigh Mercer, *Word Ways,* vol. 13, no. 1,
(February, 1980), page 36.

Leigh Mercer (1893–1977) was the Englishman who wrote the best palindrome ever: "A man a plan a canal—Panama"[6] (*Notes & Queries,* Nov. 13, 1948). Just who was this clever genius? He was an itinerant worker. In later years he became a panhandler in London who made his living by drawing

---

[6] Later someone extended it to "Zeus! A man a plan a canal—Panama? Suez?".

characters on sidewalks and hoping for donations from passers by. His sad biography can be found in *Word Ways,* vol. 24, no. 3 (August 1991), pages 131–138. The equally sad biography of Dmitri Borgmann ("the king of serious wordplay") can be found in *Word Ways,* vol. 21, no. 4 (November 1988), pages 195–198. The hobbies of these men took over their lives.

The Terms of Mathematics

I. Pure mathematics

1. *Similarly:* Two or three lines in this proof are the same as those in the previous proof.
2. *Without loss of generality:* The author proved only the easiest case.
3. *Clearly:* The author can show it in half a day.
4. *Obviously:* The author can show it in three pages.
5. *Really obvious:* It has an easy-to-understand, but hard-to-find solution.
6. *Trivial:* It has two easy-to-understand, but hard-to-find solutions.
5. *Details omitted:* The author can't quite show it.
6. *In general:* The author couldn't find his way around the pathological exceptions.
7. *It is not difficult:* It is very difficult.
8. *It is easily seen that:* Two Bell Lab scientists took $3\frac{1}{2}$ years to show it.
9. *A simple computation:* The Cray super-computer did it in under seven hours.
10. *Ingenious proof:* A proof that the author could understand. In later papers this proof will be referred to as trivial.
11. *A well-known result:* A result whose reference the author could not locate.
12. *It can be observed that:* The author hopes you haven't noticed.
13. *The author wishes to thank the referee:* The referee's name should appear as the first author.
14. *Interesting:* Dull.

This list was inspired by H. Pétard (Ralph Boas), "A Brief Dictionary of Phrases Used in Mathematical Writing," *American Mathematical Monthly,* vol. 73 (February 1966), page 196. The funny name H. Pétard (coined by Boas around 1938) has appeared in many journals over the years. A "petard" was an explosive device used in medieval times. Petards were notorious for exploding in the hands of their engineers. Anyone who is "hoisted by their own petard" is undone by their own plans.[7] Later it was said that Hector

---

[7] An alternate (and much funnier) explanation comes from the French word *pétarade.* See any French dictionary.

Pétard married Nicholas Bourbaki's daughter, Betti. Betti numbers are numbers associated with groups. They were named for Enrico Betti (1823–1892), an Italian algebraist.

II. Applied Mathematics

1. *Of great theoretical importance:* Interesting to the author.
2. *Typical results are shown:* The best results are shown.
3. *The agreement with the predicted curve is*

      Excellent   : fair
      Good   : poor
      Satisfactory   : doubtful
      Fair   : imaginary.

4. *It is suggested that* or *It is believed that* or *It is indicated that:* The author believes.
5. *It is generally believed that:* A couple of other guys think so, too.
6. *It might be argued that:* I have such a good answer to this objection that I shall now raise it.
7. *It is clear that much additional work will be required before a complete understanding is found:* The author does not understand it.

   —C.D. Graham, in *A Random Walk in Science* (Crane Russak, 1973), pages 120–121.

III. Mathematical Administration

1. *Confidence interval:* The hopeful waiting period before a request is turned down.
2. *Correlation:* An individual who discovers he shares your views shortly after you become a member of the administration.
3. *Irrational expression:* Any comment at variance with the views of the department head. —Henry Winthrop, *Recreational Mathematics Magazine* (June 1962), pages 12–15.

IV. Teaching

1. *This is one of the best textbooks in the field:* I used it in grad school.
2. *Your test scores were fairly good:* A few people managed a B.
3. *Some of you could have done better:* Everyone failed.
4. *The answer to your question is beyond the scope of this class:* I don't know.
5. *See me during office hours for a complete answer to your question:* I don't know.
6. *We will have to continue this discussion outside of class:* I thought I knew, but I don't know.

7. *Just try to understand the main ideas involved:* I couldn't follow the details either.

8. *This is a very important topic:* I understand it.

—Based on the photocopy *What the Oceanography Professor Really Means* (1994), attributed to J. Timothy Petersik.

V. Quotations

When the American astronomer, Nathaniel Bowditch, translated Laplace's treatise into English, he remarked, "I never come across one of Laplace's 'Thus it plainly appears' without feeling sure that I have hours of hard work before me to fill up the chasm and find out and show how it plainly appears." —Howard Eves, *An Introduction to the History of Mathematics, 6th ed.* (Saunders College Publishing, 1990), page 447.

Biot [Jean Baptiste Biot, 1774–1862] who assisted Laplace in revising [*Mécanique Céleste,* five volumes] for the press, says that Laplace himself was frequently unable to recover the details in the chain of reasoning, and, if satisfied that the conclusions were correct, he was content to insert the constantly recurring formula, "Il est aisé à voir" ["It is easy to see."] —W.W. Rouse Ball, *A Short Account of the History of Mathematics,* 4th ed. (1908, Dover reissue 1960), page 417.

Not seldom did [Lord Kelvin] in his writings set down some mathematical statement with the prefacing remark "It is obvious that" to the perplexity of mathematical readers to whom the statement was anything but obvious ... —S.P. Thompson, *Life of Lord Kelvin* (London, 1910), page 1136.

It was said of Jordan's writings that if he had four things on the same footing (as a, b, c, d) they would appear as a, M'_3, ϵ_2, $\Pi''_{1,2}$. — J.E. Littlewood, *Littlewood's Miscellany* (Cambridge, 1986), page 60.

It looked like a beautiful theorem—One that would place my name among the immortals. But I couldn't find a proof. I decided to lower my sights by proving a special case. I still couldn't find one I could prove. It occurred to me that by generalizing the theorem, I'd have a stronger induction hypothesis to work with. But I couldn't perform the induction. Now I'm looking for a counterexample. —David Silverman, *Journal of Recreational Mathematics,* vol. 3 (1979), page 200.

Notes

1. Lars Valerian Ahlfors was born in Helsinki in 1907. He received a Fields medal in 1936 and accepted a permanent position at Harvard in 1946. His basic area of research has been complex analysis. In 1953 he published what for many years was the standard graduate level textbook for analytic functions of one variable in complex analysis: *Complex Analysis* (248 pages). The third edition was issued in 1972. Incidentally, Mike Brill used to bounce back and forth from being an optical physicist to being an acoustical physicist. "A wave's a wave," he would say.

2. The mountain climber and the chicken-banana question are treats for students studying the cross (= vector = outer) product for the first time. Unfortunately, I have no humor concerning the dot (= scalar = inner) product, nor for the triple (= box) product. The terms *inner* and *outer* were originally chosen for what are now obscure considerations. See Michael J. Crowe, *A History of Vector Analysis* (Dover, 1985).

> [W]e distinguish between *inner* or *scalar multiplication* [$\mathbf{A} \cdot \mathbf{B}$] and *outer* or *vector multiplication* [$\mathbf{A} \times \mathbf{B}$]. Indeed, in each of these, the important property called *the distributive law of multiplication with respect to addition* . . . is valid.
>
> Why this language of vector analysis has been so firmly adopted I am unable to fully understand. It well may have some connection with the fact that many people derive much pleasure from such formal analogies with the common time-honored operations of reckoning. In any event these names for the vector operations have been accepted with tolerable generality. —Felix Klein (1925), *Geometry* (Dover, 1939), pages 50–51.

3. Max Zorn (1906–1993) published his lemma (also attributed to Hausdorff and Kuratowski) while at Yale (1934–1936). A biography of Zorn appears in the June 1993 issue of the Mathematical Association of America's *FOCUS*. A very touching tribute to his life (written by his grandson) appeared in the October 1993 (vol. 66, no. 4) issue of *Mathematics Magazine*.

4. ... it's very important that a function doesn't have to be continuous, and that a continuous function doesn't have to be differentiable, that a differentiable function doesn't have to be twice differentiable, and so on; that even if a function has derivatives of all orders, the Taylor series for this function isn't necessarily convergent, and that even if it is, its sum doesn't necessarily coincide with the value of the function [unless the remainder converges to zero]! If this coincidence takes place, the function is called analytic, and this class of functions (so the devotees of real variable function theory maintained) is so narrow that it lies outside the bounds of mainstream mathematics. And these were the only functions I'd been looking at! —I. M. Gelfand, *Quantum* (January/February 1991), page 25.

10
My Favorite Proofs

A missionary was captured by a tribe of cannibals and told that he would be boiled in oil and eaten on the following day. While in a guarded hut and waiting for his execution, he happened to notice an almanac that had been left by the previous occupant. A casual reading of the almanac showed the missionary an extraordinary coincidence: On the next day at exactly noon at exactly his location, there was going to be a total eclipse of the sun. The missionary threw down the almanac, jumped to his feet, and approached the guard.

"Excuse me, but at what time tomorrow will I be put into the pot of boiling water?"

The Guard responded, "At 12:10."

"Great," said the missionary.

"Right after the eclipse," responded the guard. "Why do you ask?"[1]

A version of this story continues: After the missionary regained his composure, he asked how the members of a primitive tribe knew about the forthcoming eclipse. The guard responded that, although they were primitive, mathematics had been highly cultivated and much appreciated in their society. In fact, there was a rarely-used clause in the tribal code that allowed any captive to reclaim his life if he could offer an elegant proof to an old theorem, or

---

[1] The missionary was planning to say, "You had better not kill me, or the gods will put out the sun. Try it and you'll see." Then shortly before he was to be put into the pot of boiling oil, the sun would start to dim due to the eclipse. The horrified primitives would recant and treat the stranger as one of the gods. This trick was employed in H. Rider Haggard's *King Solomon's Mines* (1885) and in Mark Twain's *A Connecticut Yankee in King Arthur's Court* (1889).

prove a new and interesting theorem. This joke gives me an opportunity to show off some of my favorite proofs.

Theorem 1. i^i is a real number.

Proof (direct).   The first step is the statement of Euler's formula, which is often given as a definition of $e^{i\theta}$.

$$e^{i\theta} = \cos\theta + i\sin\theta$$

$$e^{i\pi/2} = \cos\frac{\pi}{2} + i\sin\frac{\pi}{2}$$

$$e^{i\pi/2} = i$$

$$\left(e^{i\pi/2}\right)^i = i^i$$

$$e^{-\pi/2} = i^i$$

$$0.20287\ldots = i^i \qquad \text{Q.E.D.}$$

Remark. This fact inspired Professor Nicholas J. Rose to compose the following rhyme:

> The i-th power of i seems mysterious
> But if you take it quite serious,
>   You can show the following is true:
>   It's e to the minus π over 2.
> You have to admit that's quite curious.
> —N.J. Rose, *Mathematical Maxims and Minims*
> (Rome Press, 1988), page 109.

Theorem 2. $\sqrt{2}$ is irrational.

Proof (indirect).   If $\sqrt{2}$ is rational, then there must exist a smallest positive integer q, so that $\sqrt{2}q$ is an integer, but this is contradicted by the fact that $k = (\sqrt{2} - 1)q$ is a *smaller* positive integer with the same property. Q.E.D.

Remark. Most readers have seen Aristotle's proof that $\sqrt{2}$ is irrational. In his book *A Mathematician's Apology* (Cambridge, 1940), G.H. Hardy stated that Aristotle's proof was "of the highest class." That proof took a page in Hardy's little book. But the one-sentence proof given here is a simpler (and I think a prettier) proof of the same fact. Hardy also included Euclid's proof of the infinitude of prime numbers as his second example of a high-class proof.

Theorem 3.   An irrational number raised to the power of another irrational number may be rational.

Proof (non-constructive).   Consider $x = \sqrt{2}^{\sqrt{2}}$. If x is rational, then we are done. If x is not rational, then because $x^{\sqrt{2}} = 2$ is rational we are done. Q.E.D.

Remark. We still have no example of an irrational number raised to the power of an irrational number that yields a rational number. Yet we have shown one exists. This is the best example I know of a non-constructive proof.

Theorem 4.   There exists a one-to-one function from the half-open unit interval $[0, 1)$ onto the closed unit interval $[0, 1]$. In other words, it is possible to remove the end point of an interval and rearrange the numbers to fill the gap without creating a new gap. [Try to generate such a function before reading on. Things that are very easy to understand seem easy to discover, but this is an illusion. Most of the people to whom I have shown this theorem could not prove it.]

Proof (by example).   Consider the function

$$f(x) = \begin{cases} 2x, & \text{if } x = \frac{1}{2^n} \\ x, & \text{otherwise.} \end{cases}$$

This maps $1/2$ to 1, $1/4$ to $1/2$, $1/8$ to $1/4$, etc. Since there is an infinite number of rationals, no gap is created. Q.E.D.

Theorem 5.   The shortest distance from point A to line L and back to point B is the path of a mirror reflection (the path that makes equal angles with the line).

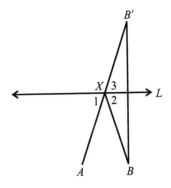

FIGURE 18

Construction. Reflect point B across line L to B' so that line L is the perpendicular bisector of segment BB'. Let X be the intersection of AB' with line L. I claim A-X-B is the minimal path from A to L to B and that $\angle 1 = \angle 2$.

Proof (Heron, circa A.D. 75). Choose any point P ($\neq X$) on L. Then (by the triangle inequality):

$$AX + XB = AX + XB' = AB' < AP + PB' = AP + PB.$$

Since $\angle 1 = \angle 3$ (vertical angles) and $\angle 3 = \angle 2$ (congruent triangles) we have $\angle 1 = \angle 2$. Q.E.D.

Theorem 7. The Law of Sines.

Proof. Consider what I call the aha formula

$$a\, h/a = b\, h/b. \tag{1}$$

Since this is an identity over the positive numbers, I can draw any figure, arbitrarily label any sides a, b, h, and the formula will describe the figure. I choose a triangle with sides a and b having the same height (h) from the base (see figure 19). Now $h/a = \sin B$ and $h/b = \sin A$. Thus, we have the Law of Sines: $a/\sin A = b/\sin B$. (The case where $\angle A$ or $\angle B$ is obtuse is dispatched by recalling $\sin x = \sin(\pi - x)$.) Q.E.D.

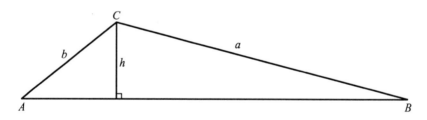

FIGURE 19

Application. Now consider a perfectly flexible cord of uniform shape and density draped over the triangle from A to C to B (here we hold $\angle A$ and $\angle B$ *never* to be obtuse). The weight of that part of the cord from A to C is b weight units; the length of that part of the cord is b linear units. The weight of that part of the cord from B to C is a weight units; the length of that part of the cord is a linear units. The Law of Sines (in the aha form) seems to say that the cord should be in equilibrium. For example, if the cord

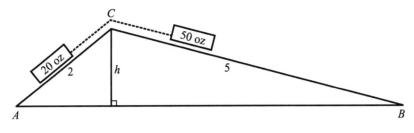

FIGURE 20

weighs 10 oz per inch and a is 5 inches and b is 2 inches, then the *aha* formula gives us $5(10) \times h/5 = 2(10) \times h/2$, and (generalizing) the two weights in Figure 20 should balance. But does gravity really act this way? Will the weights (or equivalently the flexible frictionless cord) really balance? The question was settled in 1605 by Simon Stevin (SEE·mawn stuh·VINE) a Dutch-Flemish engineer and mathematician who suggested that the cord be made endless (Figure 21). The looped cord of Figure 21 must be static (else perpetual motion). Since the base is level, the bottom portion of the cord is symmetrical and tugs equally on points A and B. Hence, removal of the bottom section will have no effect, and afterwards the top portion will still remain in equilibrium. By this clever trick, Stevin proved the "principle of the inclined plane,"[2] which is described mathematically by the Law of Sines.

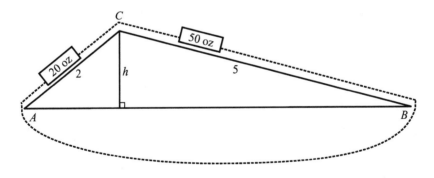

FIGURE 21

---

[2] The principle of the inclined plane states: On inclined planes of equal height (from a level base) equal weights tug in inverse proportion to the lengths of their planes. This means that if we double the length of the plane, then the same object will tug with only half the force it had before the plane was doubled. In Figure 19, if $x/a = y/b$, then a weight of x oz sliding down a plane of length a in will balance with a weight of y oz sliding down a plane of length b in.

Remark. Stevin placed Figure 21 on the title page of his 1605 book and later had it inscribed on his tomb with the words (in translation) "that which is wonderful is not so wonderful" meaning, I think, that once amazing things are explained they often appear so obvious as to lose their charm. If that is the case here, then we are more than compensated by the charm of Stevin's thinking.

These proofs are part of why I love mathematics. They are more than clever—they are ingenious. No detective story was ever this smart. No painting was ever this beautiful. Nothing I have personally experienced has been so compelling. There has been much discussion about how to interest young people in mathematics. To the few who resonate with the subject, there is no keeping them away from mathematics. The study of mathematics is its own reward.

"Farewell, farewell! but this I tell
To thee, thou Wedding-Guest!
He prayeth well, who loveth well
Both man and bird and beast.

He prayeth best, who loveth best
All things both great and small;
For the dear God who loveth us,
He made and loveth all."

   —Samuel Taylor Coleridge,
    The Rime of the Ancient Mariner (1834)

Index